EAI/Springer Innovations in Communication and Computing

Series Editor

Imrich Chlamtac, European Alliance for Innovation, Ghent, Belgium

The impact of information technologies is creating a new world yet not fully understood. The extent and speed of economic, life style and social changes already perceived in everyday life is hard to estimate without understanding the technological driving forces behind it. This series presents contributed volumes featuring the latest research and development in the various information engineering technologies that play a key role in this process. The range of topics, focusing primarily on communications and computing engineering include, but are not limited to, wireless networks; mobile communication; design and learning; gaming; interaction; e-health and pervasive healthcare; energy management; smart grids; internet of things; cognitive radio networks; computation; cloud computing; ubiquitous connectivity, and in mode general smart living, smart cities, Internet of Things and more. The series publishes a combination of expanded papers selected from hosted and sponsored European Alliance for Innovation (EAI) conferences that present cutting edge, global research as well as provide new perspectives on traditional related engineering fields. This content, complemented with open calls for contribution of book titles and individual chapters, together maintain Springer's and EAI's high standards of academic excellence. The audience for the books consists of researchers, industry professionals, advanced level students as well as practitioners in related fields of activity include information and communication specialists, security experts, economists, urban planners, doctors, and in general representatives in all those walks of life affected ad contributing to the information revolution.

Indexing: This series is indexed in Scopus, Ei Compendex, and zbMATH.

About EAI - EAI is a grassroots member organization initiated through cooperation between businesses, public, private and government organizations to address the global challenges of Europe's future competitiveness and link the European Research community with its counterparts around the globe. EAI reaches out to hundreds of thousands of individual subscribers on all continents and collaborates with an institutional member base including Fortune 500 companies, government organizations, and educational institutions, provide a free research and innovation platform. Through its open free membership model EAI promotes a new research and innovation culture based on collaboration, connectivity and recognition of excellence by community.

Lucia Knapčíková • Dragan Peraković
Editors

8th EAI International Conference on Management of Manufacturing Systems

Springer

EAI
RESEARCH MEETS INNOVATION

Editors
Lucia Knapčíková
Faculty of Manufacturing Technologies
Technical University of Košice
Prešov, Slovakia

Dragan Peraković
Faculty of Transport and Traffic Sciences
University of Zagreb
Zagreb, Croatia

ISSN 2522-8595 ISSN 2522-8609 (electronic)
EAI/Springer Innovations in Communication and Computing
ISBN 978-3-031-53160-6 ISBN 978-3-031-53161-3 (eBook)
https://doi.org/10.1007/978-3-031-53161-3

This Springer imprint is published by the registered company Springer Nature Switzerland AG
The registered company address is: Gewerbestrasse 11, 6330 Cham, Switzerland

Credits for Organizing Committee

EAI MMS 2023 – 8th EAI International Conference on Management of Manufacturing Systems

October 24–26, 2023
Bratislava, Slovakia

Steering Committee

Chair
Prof. Imrich Chlamtac
Bruno Kessler Professor, University of Trento, Italy

Member
Assoc. Prof. Lucia Knapčíková, Ph.D.
Technical University of Košice, Faculty of Manufacturing Technologies, Slovak Republic
Prof. Dragan Peraković, Ph.D.
University of Zagreb, Faculty of Transport and Traffic Science, Croatia

Organizing Committee

General Chair
Assoc. Prof. Dr. Lucia Knapčíková
Technical University of Košice, Slovak Republic

General Co-Chair
Prof. Dr. Dragan Peraković
University of Zagreb, Faculty of Transport and Traffic Science, Croatia

Technical Program Committee Chairs
Assoc. Prof. Dr. Lucia Knapčíková
Technical University of Košice, Slovak Republic
Prof. Dr. Dragan Peraković
University of Zagreb, Faculty of Transport and Traffic Science, Croatia

Technical Program Committee Co-Chair
Prof. Franz Haas
Graz University of Technology, Austria
Assist. Prof. Dr. Annamaria Behúnová
Technical University of Košice, Slovak Republic
Assist. Prof. Dr. Ivan Cvitić
University of Zagreb, Faculty of Transport and Traffic Science, Croatia

Web Chair
Dr. Jozef Husár
Technical University of Košice, Slovak Republic

Publicity and Social Media Chair
Assoc. Prof. Dr. Lucia Knapčíková
Technical University of Košice, Slovak Republic
Prof. Dr. Dragan Peraković
University of Zagreb, Faculty of Transport and Traffic Science, Croatia

Workshop Chair
Dr. Jozef Husár
Technical University of Košice, Slovak Republic

Sponsorship & Exhibit Chair
Assoc. Prof. Dr. Lucia Knapčíková
Technical University of Košice, Slovak Republic
Prof. Dr. Dragan Peraković
University of Zagreb, Faculty of Transport and Traffic Science, Croatia

Publication Chair
Assoc. Prof. Dr. Lucia Knapčíková
Technical University of Košice, Slovak Republic
Prof. Dr. Dragan Peraković
University of Zagreb, Faculty of Transport and Traffic Science, Croatia

Panel Chair
Prof. Dr. Dragan Peraković
University of Zagreb, Faculty of Transport and Traffic Science, Croatia

Local Chair
Assoc. Prof. Dr. Lucia Knapčíková
Technical University of Kosice, Faculty of Manufacturing Technologies, Slovakia

Preface

The year 2023 is special because the 8th International Conference on Management of Manufacturing Systems (EAI MMS 2023) was part of the prestigious Smart Life Summit. The summit's goal was to bring together academics, companies, stakeholders, and regional and national governments with an area of interest in Artificial Intelligence, Health, Energy, Security, Mobility, and Sustainability. The unique combination of fields and disciplines focuses on Industry 4.0, the power of smarter information, and opportunities to create a bridge between science and practice. We are delighted to introduce the eighth edition of the 2023 European Alliance for Innovation (EAI) International Conference on Management of Manufacturing Systems. This conference has brought researchers, developers, and practitioners around the world. In light of the latest knowledge and findings from scientific projects, the authors present actual R&D trends in the given field. This issue defines the state of the art in the area and explores related topics for future research. The proceeding of EAI MMS 2023 consisted of 12 accepted full papers in oral presentation sessions at the main conference tracks. The keynote speaker was Mr. Nathanael Dadon from EV Meter, Ltd., Israel, with a Keynote Speech, "The main challenges of electric mobility – charging stations and energy management". Working with such an excellent organizing committee team was also a great pleasure. We are also grateful to Conference EAI Manager Radka Vasileiadis and Katarina Antalova, and Eliska Vlckova – Managing Editor, for their support, all the authors who submitted their papers to the EAI MMS 2023 conference, and reviewers for their hard work. The ambition of the conference was to establish communication channels and disseminate knowledge among professionals working in manufacturing and related institutions. We strongly believe that future cooperation will be as successful and stimulating as indicated by the contributions presented in this volume.

Prešov, Slovakia
Zagreb, Croatia
Bratislava, Slovak Republic
24–26 October 2023

Lucia Knapčíková
Dragan Peraković

Contents

Part I
Smart Techniques/Smart Technology

Input–Output Product Model Adapted to Material Flow in Order to Achieve Cleaner Production in Industrial Enterprises

Matouš Machka, Pavel Scholz, and Theodor Beran

1 Introduction

The product input–output model is intended for matrix costing, and, together with other input–output models, represents one of the main tools of input–output analysis. A major contribution to the development of this discipline has been made by the Russian-American economist and the holder of Nobel Prize in economy, Wassily Leontief (1905–1999). In his best-known book, *Structure of the American Economy* 1919–1929, Leontief provides a thorough analysis of inter-industrial relations within the context of the US economy of 1939. The analysis is further complemented by a thorough mathematical description of the open static input–output model [8]. Each input–output model comprises a set of elements that create a specific economic system. The open input–output model is only used in cases where the mutual supplier–customer relationships of individual elements constituting the model can be defined [7]. The input–output model builds upon the mathematical foundations based on the microeconomic theory of production function. This theory can be used to thoroughly describe the mutual economic cooperation between the different elements of the model in the form of mutual provision of supplies, goods, services, and products. Open input–output models are divided into two types, the organizational type and the technological type. The organizational input–output model can be a powerful tool for the needs of short-term planning of supply, production, and distribution of goods, products, and services. Organizational model elements may be defined as cost centers of given enterprise, industrial enterprises, or entire branches of economy that cooperate with one another through the supply of goods, products, and services. In general, it can be assumed that an input–output

M. Machka (✉) · P. Scholz · T. Beran
Czech Technical University in Prague, Prague, Czech Republic
e-mail: matous.machka@fs.cvut.cz

L. Knapčíková, D. Peraković (eds.), *8th EAI International Conference on Management of Manufacturing Systems*, EAI/Springer Innovations in Communication and Computing, https://doi.org/10.1007/978-3-031-53161-3_1

model based on cost centers represents the lowest level of organizational input–output models. The elements of a technological model represent the components constituting a product, groups of similar products, or entire branches dedicated to production of similar groups of products. The input–output product model represents the lowest level of technological input–output models. Due to the interdependence of individual elements, it is possible to aggregate individual input–output models independently of their type and level. Aggregation can be defined as integration of organizational or technological input–output models into one aggregated input–output model. The integration of aggregated organizational and technological input–output models enables us to create a model describing the chosen sector of national economy [7, 8, 11], which represents an important tool for the needs of macroeconomic analyses. Input–output models constructed at the lowest levels can be aggregated with each other and subsequently transformed into higher level models, therefore it is very important to pay close attention to them. Industrial enterprises with complex structures which may be defined as microeconomic systems present the ideal conditions for construction and testing of such models [15]. The examined elements may represent either cost centers or manufactured products.

This paper focuses on a product input–output model adapted to material and energetic flows in an industrial enterprise. The proposed product input–output model is supposed to facilitate the process of matrix costing implementation in industrial SMEs and significantly reduce the complexity of realization of set targets within the context of negative environmental impact reduction. In particular, these impacts include reduction of energy consumption and waste volume, i.e., material losses incurred during production. The authors chose the agricultural square silo as the test subject for the proposed input–output model. The first section of this thesis is dedicated to theoretical background of the issue, including the introduction of the authors' hypothesis. The following section dedicates space to structural and technological analysis of the examined product. The next section will describe the proposed product input–output model and derivation of relevant equations needed for matrix costing itself. The last part discusses the topic of production process digitalization and its integration with product input–output model. The conclusion will present the obtained results and provide detailed explanation of those findings, including possible plans for future research.

2 Theoretical Background

During the long development of this discipline, many interesting publications have been written, focusing on the application of organizational input–output models in the secondary sector of the economy. Among the most interesting works are scientific articles dealing with implementation of these models in mining [6, 15] and manufacturing industries [1]. Other interesting works focus on the question of input–output models' utilization for the needs of effective decision-making, in order to reduce negative impacts on the environment. This issue is mainly concerned with the question of reduction of energy consumption [10, 11] and waste volume generated during production activities [2]. An input–output model that reflects these environmental impacts can provide industrial enterprise production managers with a high-quality source of information, helping to achieve production that is both cleaner and more energy-efficient. Another benefit of such models is the ability to choose a suitable provider of energy, goods, services, and supplies. Suitable provider being a provider that consumes the least amount of energy and generates the least amount of waste during their operation. If the industrial enterprise managers started using proper input–output models as the basis of their decision-making more often, their future decision-making strategies could lead to the decrease of negative environmental impacts throughout entire industrial regions. Currently, the organizational–technological input–output model adapted to the MFCA (material flow cost accounting) system represents a highly effective tool for achieving the above-mentioned environmental objectives of any given enterprise [2].

The aim of this paper is to focus on the technological part of the model mentioned above and define its connection to the product input–output model, which would simplify the process of costing of individual cost objects. Matrix costing via MFCA-adapted input–output model is a highly demanding task, and any attempt to use this approach within the context of industrial SMEs could quite possibly run into difficulties during its implementation. For this reason, the authors of this paper have established the following hypothesis.

> A product input–output model tailored to the MFCA system will facilitate the matrix costing process and provide satisfactory results for the needs of effective decision-making and fulfillment of set environmental goals in industrial SMEs.

3 Structural–Technological Analysis of a Square Silo

Implementing a product input–output model in industrial enterprises requires thorough preparation of initial data. During the first stage, it is therefore necessary to carry out a detailed structural and technological analysis of given engineering product [9]. The use of the product input–output model can be especially effective for products, components of which are characterized by structural–technological similarity [9]. Such characteristic greatly simplifies the process of structuring of parts and assembly groups in the first quadrant of the product input–output model. In the following two subsections, the authors directly build upon the previously proposed methodology of implementing these models in industrial practice [9]. Up until now, the product input–output model has only been tested under the conditions of job production of structural steel [9], but the square agricultural silo shown in Fig. 1 offers also the option of batch production. This fact could increase the potential of product input–output model utilization within the industrial sector itself. The square agricultural silo was chosen for three reasons. The first reason reflects the increasing demand for new silos of this type among agricultural enterprises, as the existing units are already at the end of their life cycle. The second reason

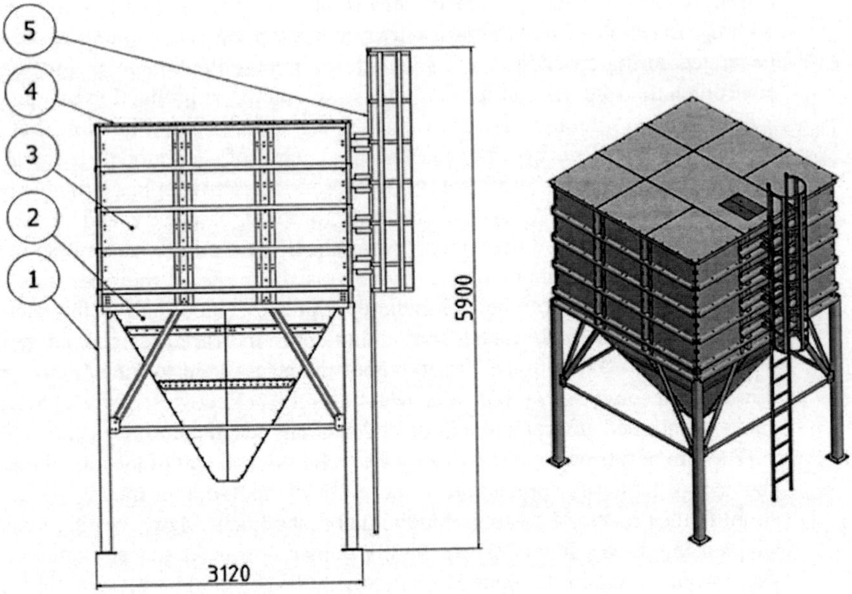

Fig. 1 Conceptual design of the square agricultural silo, where the marked assembly groups represent: 1 – Support; 2 – Funnel; 3 – Container; 4 – Roof; 5 – Ladder. (Source: own design)

lies in the technical design of components the square silos use, components which are characterized by structural–technological similarity. The third reason being the previously mentioned possibility of batch production of square silos. It offers a great opportunity to digitalize production activities and analyze them in detail. The main objective of such analysis is to determine appropriate times required for the implementation of individual production activities and test the proposed methodology. As part of the research, a digital model of a factory conforming to the standard of a medium-sized engineering industrial enterprise engaged in batch production of these silos was also created. The first subsection will be devoted to the structural analysis of the silo, in order to efficiently process the initial data for the first quadrant of the product input–output model. The second subsection briefly explains the production process used as a basis for the digital simulation of production and the appropriate choice of quantity centers within the requirements of the MFCA system.

3.1 *Structural Analysis of a Square Silo*

The bill of material (BOM) attached to the technical report represents the elementary source of initial information for the construction of the product input–output model.

Table 1 BOM structure

Description of assembly groups			Total number of	
Number	Name	Material	components	component types
5	Ladder	standardized part		
4	Roof	plain GI sheets	8	3
3	Container	plain GI sheet	80	9
2	Funnel	plain GI sheet	52	14
1	Support	plain GI sheet	40	7
1	Support	steel square tubes	4	1

Source: own analysis

Table 2 The arrangement of individual component types into classes within the requirements of the construction of the first quadrant of the product input–output model

Material	WB2	WB3	WB4	WB6	WB12	FRQ120
Class	Q_1	Q_2	Q_3	Q_4	Q_5	Q_6
Component types	$Q_{1,1}$	$Q_{2,1}$	$Q_{3,1}$	$Q_{4,1}$	$Q_{5,1}$	$Q_{6,1}$
	⋮	⋮				
	$Q_{1,5}$	⋮				
		$Q_{2,24}$				
Total types	5	24	1	1	1	1

Source: own analysis

According to Fig. 1, the square agricultural silo consists of five assembly groups, where each assembly group consists of a given number of components. Components are further classified according to material (material, according to Table 1) and their structural role within the assembly group (component types, according to Table 1). A clear structure of the bill of materials for a square agricultural silo is presented in Table 1.

The sum of the number of component types and assembly groups determines the dimension of the square matrix expressing the first quadrant of the product input–output model. As per the requirements of the creation of the first quadrant, we structure each component type into classes, based on the assignment of appropriate indices [9]. The classes represent vectors, or column matrices, and their elements are labeled with the letter Q_{ij}, where the i index indicates the type of material used and the j index indicates the specific component. The values of these variables indicate the number of selected components. The overall structure is shown in Table 2. Based on this table, it can be seen that two types of material will be required for production purposes, namely, the plain sheets marked with the WBX symbol, and the square steel tubes marked with FRX. The letter X indicates the basic dimension of given material. Thickness in the case of the plain sheets, and length of the cross-section edge in the case of the square steel tubes.

The structure of the assembly groups can be expressed mathematically by Eq. (1), which applies to all assembly groups except the safety ladder. This ladder is provided by the supplier and therefore does not enter the production process. Whenever a component is not present in selected assembly group, its element will be zero. Matrix dimensions must be the same for all assembly groups. In the case of the square agricultural silo, the dimensions will be equal to the total number of types of individual components.

$$G_{i\ 33}^{\mathrm{T}} = [Q_1, Q_2, Q_3, Q_4, Q_5, Q_6]^{\mathrm{T}} \tag{1}$$

Where

$G_{i\ 33}^{\mathrm{T}}$ ... a transposed column matrix expressing the numbers of individual component types within assembly group #i

An efficient data structure requires the use of similar markings for individual components and assembly groups, including the indicated indexation, in given CAD software. In this case, the drawing documentation will include 33 component drawings, five assembly group drawings, and the final assembly drawing. One of the assembly group drawings contains the assembly of the support beam. This subassembly is not shown in Fig. 1, since it is a part of the assembly group Support. Thus, each component created will be stored under the designation of given element, e.g., components will be designated as *Q-i-j*, while the assembly groups will be designated as *G-i*. After creating the appropriate assemblies in the CAD software, this data can be exported from the digital bill of materials to a spreadsheet software of choice, where we create the necessary matrices in accordance with Eq. (1). Selected data will include the component designation and number of pieces used in given assembly. This step represents the initial and necessary condition for creation of the first quadrant of the product input–output model.

3.2 Technological Analysis of a Square Silo

A schematic representation of the production process is shown in Fig. 2. First, we focus on the explanation of production activities related to components manufactured from plain sheets. During the first stage of production, these components must be cut using the laser-cutting center. Cut parts will then be transported to the bending department. Subsequently, they will be moved into the warehouse, except for a portion which will be moved to the soldering department. Components manufactured from steel parallelepipeds will first be cut using the NC band saw and then moved to the soldering department. The soldering department will produce support beams, which will then be moved into the warehouse. Assembly of the square silo will be carried out at a prearranged location chosen by the customer.

At this stage, it is also necessary to break down the production process and adapt it to the requirements of the MFCA system. The basic MFCA system requirement involves the formulation of quantity centers. Traditional cost accounting (TCA) systems work with cost centers. For the ease of implementation of the MFCA system, however, cost centers can be divided into quantity centers, where the efficient costs of products and the inefficient costs of material losses are recorded

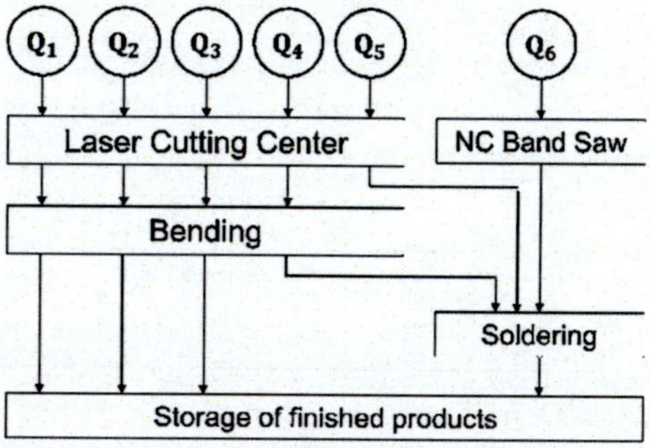

Fig. 2 Scheme of the production process for square agricultural silo. (Source: own design)

[2]. The quantity centers will represent the individual production departments shown in Fig. 2. In order to reduce negative environmental impacts during production, production managers must identify the section of production which represents the greatest source of material losses. In this case, it is the section of laser cutting of plain sheets. Production activities carried out within this section must be set up in a way, such that would maximize the number of components cut from a single plain sheet. Understandably, similar strategy must be applied to the NC band saw-cutting section.

4 Product Input–Output Model Adapted to the MFCA System

A product input–output model can be defined as a static open input–output model that represents the lowest level of technological input–output models. This model provides the basis for possible aggregation of other technological input–output models describing given branch, in this case the industrial production of agricultural silos. Technological input–output models do not change their structure during any given period, i.e., the mutual cooperation between individual elements remains the same. This distinguishes them from organizational models whose structure can change over time. These changes are often triggered by changes in the organizational structures of industrial enterprises [11, 16]. The results of matrix costing performed using the product input–output model provide a high-quality basis for pricing of individual components. The product input–output model adapted to the MFCA system will also provide important information about material losses and energy consumption during production of given component. To some extent, the implementation of these models in industrial SMEs can contribute to improving the quality of information flows within the context of in-house departments. This advantage will make it possible to minimize the risks related to the lack of information base within industrial SMEs, which still face frequent financial accounting inaccuracies [3].

i \ j	1	...	N	1	...	M	FA	Total
	Q. I. A			**Q. I. B**			**Q. II.**	
1	$q_{1,1} \cdot p_1$	...	$q_{1,N} \cdot p_1$	$q_{1,N+1} \cdot p_1$	...	$q_{1,N+M} \cdot p_1$	$f_1 \cdot p_1$	$Q_1 \cdot p_1$
⋮	⋮	⋮	⋮	⋮	⋮	⋮	⋮	⋮
N	$q_{N,1} \cdot p_N$	...	$q_{N,N} \cdot p_N$	$q_{N,N+1} \cdot p_N$	...	$q_{N,N+M} \cdot p_N$	$f_N \cdot p_N$	$Q_N \cdot p_N$
1	$q_{N+1,1} \cdot p_{N+1}$	...	$q_{N+1,N} \cdot p_{N+1}$	$q_{N+1,N+1} \cdot p_{N+1}$	...	$q_{N+1,N+M} \cdot p_{N+1}$	$f_{N+1} \cdot p_{N+1}$	$Q_{N+1} \cdot p_{N+1}$
⋮	⋮	⋮	⋮	⋮	⋮	⋮	⋮	⋮
M	$q_{N+M,1} \cdot p_{N+M}$	...	$q_{N+M,N} \cdot p_{N+M}$	$q_{N+M,N+1} \cdot p_{N+M}$		$q_{N+M,N+M} \cdot p_{N+M}$	$f_{N+M} \cdot p_{N+M}$	$Q_{N+M} \cdot p_{N+M}$
	Q. III. A			**Q. III. B**			**Total**	
Material	c_{11}	...	$c_{1,N}$	$c_{1,N+1}$	...	$c_{1,N+M}$	c_1	
Energy	c_{21}	...	$c_{2,N}$	$c_{2,N+1}$	...	$c_{2,N+M}$	c_2	
System	c_{31}	...	$c_{3,N}$	$c_{3,N+1}$	...	$c_{3,N+M}$	c_3	
Waste	c_{41}	...	$c_{4,N}$	$c_{4,N+1}$	...	$c_{4,N+M}$	c_4	
Total	$Q_1 \cdot p_1$	...	$Q_N \cdot p_N$	$Q_{N+1} \cdot p_{N+1}$	...	$Q_{N+M} \cdot p_{N+M}$		$\Sigma\Sigma$

Fig. 3 Schematic design of product input–output model adapted to the MFCA system. (Source: own design)

4.1 Mathematical Interpretation of the Model

Figure 3 illustrates the product input–output model adapted to the MFCA system. The elements of this model represent the individual components and assembly groups of given product, in this case, a square agricultural silo. Interconnections between these elements are defined using three quadrants forming an input–output model. These quadrants can be mathematically expressed in the form of matrices, which represent the basis for derivation of selected input–output modeling tasks. The meaning of these quadrants is explained in the following points.

- *The first quadrant – Q.I*: internal turnover square matrix for the examined economic system. Output elements with index *j* represent components that are created from input elements with index *i*. This quadrant consists of two parts. In this particular case, the Q.I.A part will be zero, because it comprises individual components. The Q.I.B part represents assembly groups (output elements) consisting of components (input elements). Exact structure of this part of the quadrant can be obtained by transforming the data from the digital assembly bill of materials of individual assembly groups.
- *The second quadrant – Q.II*: final use vector, in this case, the final assembly vector. This quadrant records the number of components and directly reflects the values obtained from the digital bill of materials of final assembly.
- *The third quadrant – Q.III*: quadrant of primary costs incurred by production of individual components in quantity centers. Quantity centers' costs are calculated by category and not by cost types [2], in order to provide a good record of material losses and energy consumption during production.

Matrix costing is based on a modified version of the third input–output modeling task, which is based on the cost Eq. (2) [11].

$$Q_j \cdot p_j = \sum_{i=1}^{N} q_{ij} \cdot p_i + \sum_{i=N+1}^{N+M} q_{ij} \cdot p_i + \sum_{i=1}^{4} c_{ij} \quad (2)$$

Where

Q_j ... total number of pieces of the *j*-component
p_j ... an unknown price of a *j*-component
q_{ij} ... number of pieces of the input *i*-component in the output *j*-component
c_{ij} ... *i*-category of the cost required to implement the output *j*-component
N ... number of individual component types
M ... number of assembly groups

Equation (2) is further linearized using the Leontief production function (3). The linearized equation is written into a system of linear Eq. (4) and then converted into matrix form (5).

$$q_{ij} = \widetilde{a_{ij}} \cdot Q_i \tag{3}$$

Where

$\widetilde{a_{ij}}$ ... technical coefficient related to the number of pieces of the input *i*-component in the output *j*-component

$$\begin{aligned} Q_j \cdot p_j = \widetilde{a_{1j}} \cdot Q_1 \cdot p_1 + \cdots + \widetilde{a_{Nj}} \cdot Q_N \cdot p_N + \cdots + \widetilde{a_{N+1,j}} \cdot Q_{N+1} \cdot p_{N+1} + \ldots \\ + \widetilde{a_{N+M,j}} \cdot Q_{N+M} \cdot p_{N+M} + c_{11} + \cdots + c_{4j} \end{aligned} \tag{4}$$

$$Q_n = \widetilde{\boldsymbol{A}}^{\mathrm{T}}_{n \times n} \cdot Q_n + C_n \tag{5}$$

Where

$Q_n = [Q_i \cdot p_i]$ ... vector of production costs
$\widetilde{\boldsymbol{A}}^{\mathrm{T}}_{n \times n}$ ... transposed square matrix of technical coefficients
C_n ... vector of cost categories

After expressing the vector of production costs from Eq. (5), we obtain the price Eq. (6), which represents the third input–output modeling equation.

$$Q_n = \left[\boldsymbol{E}_{n \times n} - \widetilde{\boldsymbol{A}}^{\mathrm{T}}_{n \times n} \right]^{-1} \cdot C_n = \widetilde{\boldsymbol{B}}_{n \times n} \cdot C_n \tag{6}$$

Where

$\boldsymbol{E}_{n \times n}$ ... identity matrix
$\widetilde{\boldsymbol{B}}_{n \times n}$ ... matrix of complex consumption coefficients

The unit costs of each component are obtained by dividing the elements of the production cost vector by the respective numbers of given components (7).

$$P_n = [p_i] = \left[\frac{Q_i \cdot p_i}{Q_i}\right] \tag{7}$$

Using the classical expression of the balance Eq. (8), we derive the relationship needed for calculation of rejected product coefficient used in the organizational–technological input–output model adapted to the MFCA system [2]. Accurate calculation can only be made using the relationship derived from a product input–output model.

$$Q_i \cdot p_i = \sum_{j=1}^{N} q_{ij} \cdot p_i + \sum_{j=N+1}^{N+M} q_{ij} \cdot p_i + f_i \cdot p_i \tag{8}$$

Where

$f_i \ldots$ number of components entering the final assembly

Dividing Eq. (8) by the price coefficient yields the balance equation expressed only in terms of the number of pieces of individual components (9). Given numbers of pieces must be in agreement with the digital bill of materials of the examined product.

$$Q_i = \sum_{j=1}^{N} q_{ij} + \sum_{j=N+1}^{N+M} q_{ij} + f_i \tag{9}$$

Within the context of MFCA system conditions, it can be assumed that individual variables express the sum of good and defective pieces, i.e., rejects. This assumption is expressed by Eq. (10).

$$Q_i' + R_i = \sum_{j=1}^{N} \left(q_{ij}' + r_{ij}\right) + \sum_{j=N+1}^{N+M} \left(q_{ij}' + r_{ij}\right) + f_i' + r_i \tag{10}$$

The preceding equation can be broken down into two equations, namely, the balance equation determining the number of good pieces (11) and the balance equation associated with rejected pieces (12), denoted by letters R and r.

$$Q_i' = \sum_{j=1}^{N} q_{ij}' + \sum_{j=N+1}^{N+M} q_{ij}' + f_i' \tag{11}$$

$$R_i = \sum_{j=1}^{N} r_{ij} + \sum_{j=N+1}^{N+M} r_{ij} + r_i \tag{12}$$

The coefficient of defective pieces β is given by the total number of defective pieces divided by the total number of component pieces (13). The precise determination of this coefficient can be considered a highly significant extension of existing understanding of the organizational–technological input–output model adapted to the MFCA system.

$$\beta_i = \frac{R_i}{Q_i} \tag{13}$$

Using linear programming, it is possible to perform an input–output analysis simulating various scenarios [16] that will include the issue of reduction of energy consumption and waste volume during the production process. Based on these possible scenarios, the authors ask the following question.

> In what way will these changes influence the initial costs and prices of individual components?

4.2 Economic Interpretation of the Model

Although matrix costing performed using the organizational–technological input–output model adapted to the MFCA system is higher in complexity, it provides information about costs associated with material flows in much greater detail. Unfortunately, matrix costing using a technological input–output model adapted to the MFCA system cannot provide such information. This is due to the technological structure of its elements, which are represented by the individual components themselves. For this reason, we need to integrate it with MFCA-adapted TCAs. For easy integration of MFCA into TCA, let us assume that these centers build upon an existing structure of cost centers, which typically consist of multiple quantity centers [2]. The cost budget will therefore be implemented at the level of designated quantity centers [2]. This gives us highly precise information about efficient and inefficient costs in any given quantity center, i.e., the cost of realized products and material losses [2]. Quantity centers were identified by technological analysis.

In each quantity center, four cost categories are recorded [2]:

- Material costs: material consumption during production.
- Energy costs: energy consumption during production.
- System costs, i.e., costs incurred during realization of products in given quantity center.
- Production waste handling costs (does not fall under the effective cost of realized products section).

Primary costs are expressed in the cost types defined in the fifth account class of the chart of account. These costs are further transformed into the four categories mentioned above and then divided into two sections [2], namely, efficient costs of products and inefficient costs of material losses. Transformation must be carried out for each quantity center [2].

For the sake of simplicity, the calculation will focus only on the allocation of the relevant cost categories. The allocation of cost categories will be carried out using an absorption costing using modern time allocation bases (15). The other three categories will be subject to cost allocation in addition to material costs, as they cannot be directly assigned to cost objects. These categories can be defined as fixed costs in this case. The allocation rate will be determined by Eq. (14).

$$\widetilde{c_j} = \frac{C_j}{T_j} \tag{14}$$

Where

$\widetilde{c_j}$ ... hourly costs rate for the *j*-technology section.
C_j ... summarization of three allocated costs category determined for the *j*-quantity center.
T_j ... time capacity for the *j*-quantity center.

$$\widetilde{c_{ij}} = t_{ij} \cdot \widetilde{c_j} \tag{15}$$

Where

$\widetilde{c_{ij}}$ ... allocated costs category for the *i*-cost unit and *j*-quantity center.
t_{ij} ... production times related to the *i*-cost unit and *j*-quantity center.

After performing the cost allocation, it will be possible to fill out the third quadrant using the calculated values and start the input–output analysis. However, it is important to recall that allocated costs encapsulate the sums of three cost categories. This efficient method of allocation offers high-quality guidance for industrial SMEs, which still have problems implementing some of the more demanding methods of calculating unit costs [4, 5].

5 Digital Simulation of the Production Process

The main goal of digital simulation of the production process of the square agricultural silo was precise identification of production times in associated quantity centers. With the help of the factory simulation model, it is possible to record the load of individual machines and workplaces with great precision, and thus also contribute to the creation of high-quality database intended for the analysis of life cycle of the factory. Factory life cycle issues are currently not satisfactorily resolved [14].

The digital production simulation was set up to produce 25 square silos in 23 days, with the quantity centers under full load. Table 3 provides the arrangement of initial data for the start of the digital simulation. First, the production material was divided into classes, and then the number of components to be produced from one piece of material was determined. Furthermore, the number of pieces of the categories that are needed for the implementation of the specified number of products was determined. Class WB represents galvanized plain sheets and class FRQ represents square support beams. Finally, the quantity of pallets on which the production material will be placed and transported was determined.

Table 3 Arrangement of data to start a production simulation

Class	Description of component types	Number of		
		parts	components	pallets
WB2N01	Roof desks	25	1	1
WB2N02	Roof desks	50	2	1
WB2N03	Roof desks	50	2	1
WB2N04	Walls of container	25	2	1
WB2N05	Walls of container	44	2	4
WB3N01	Long beams	25	4	1
WB3N02	Support consoles	25	60	1
WB3N03	Walls of funnels	50	4	1
WB3N04	Beams	25	8	3
WB3N05	Short beams	25	10	2
WB3N06	Walls of funnels	50	2	2
WB3N07	Long beams	38	4	2
WB4N01	Support walls	25	2	2
WB6N01	Strong consoles	25	12	1
WB12N01	Columns footings	25	4	1
FRQ120	Support columns	100	1	0

Source: own analysis

The values shown in Table 3 were determined using a calculation that was performed on the basis of two information sources. The first source was the bill of material itself, indicating the number of individual components for the production of one square silo. Multiplying these values by 25 gives the total number of individual components needed to produce 25 square silos. Drawing documentation was used as the second source of information in order to precisely define the sizes of the given components. Using the second source, it was determined how many components can be obtained from one piece of production material. The number of pieces of production material was set depending on the total number of individual components. Before starting the simulation of the production process, it was necessary to define the so-called Flow Items and Fixed Resources. In this case, Flow Items represent the production material (plain GI sheet and steel square tubes), individual components, and pallets. Fixed Resources are represented by production machines and warehouses. Production workers and forklift belong to the so-called Task Executers. The numerical structure of Flow Items is shown in Table 3 and a schematic representation of their flow is shown in Fig. 2. The arrangement of production machines is shown in Fig. 4, where each number represents one quantity center. The production times for individual Fixed Resources were set in such a way as to approximate the reality as much as possible. The results of the simulation of the production process in the form of time periods expressing the processing time of the production material and individual components are shown in Table 4. For each quantity center, it is also possible to create digital indicators designed to monitor the amount of waste and defective pieces that were created. This benefit facilitates

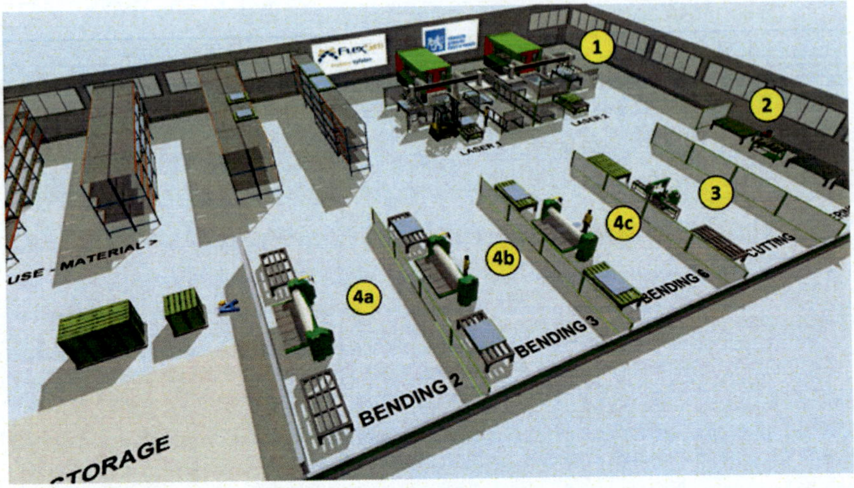

Fig. 4 Created simulation model of batch production of square agricultural silos (1 – CNC laser; 2 – soldering; 3 – NC saw; 4a, 4b and 4c – bending). (Source: own 3D simulation model designed in FlexSim software)

the optimization of production conditions because, with the help of the created indicators, it is possible to immediately identify quantity centers with a significantly higher rate of waste and defective pieces. In this way, the volume of waste can be gradually reduced and significantly cleaner production can be achieved.

The aim of the simulation was to obtain the times required for the production of given components in individual quantity centers. The results obtained are documented in Table 4. Each individual class shown in this table represents a cost object (15). Unit costs are obtained by process costing. These costs are then assigned to the third quadrant of the product input–output model to their respective components and assembly groups. The choice of total time capacity can be estimated.

Advantages of digital production process simulation for the needs of accurate costing:

- It is not necessary to undergo an expensive and time-consuming process of physical measurement of analyzed activities [12].
- Ability to efficiently monitor and record material flows between quantity centers.
- Rapid optimization of production activities within the context of production process requirements [13].
- Integration of probabilistic models into the simulation model allows accurate analysis of the place and time of defective products origin.
- Highly accurate identification of times that represent a necessary initial condition for cost allocation.
- The possibility of creating waste indicators in individual cost centers.

Table 4 Initial data obtained from digital simulation of the production process

	Quantity centers (hours)			
Class of the material	Laser	Bending	NC Saw	Soldering
FRQ120			6	31.7
FRQ120			6	31.6
FRQ120			6	31.6
FRQ120			6	31.6
WB2N01	25.9	7.7		
WB2N02	36.8	16.6		
WB2N03	31.1	16.6		
WB2N04	42.4	13.8		
WB2N05	35.5	23.4		
WB2N05	35.4	23.7		
WB2N05	37.8	23.4		
WB2N05	35.3	23.4		
WB3N01	33	19.8		
WB3N02	126	194		
WB3N03	15	16.6		
WB3N04	16.3	33.5		
WB3N04	18.4	33		
WB3N04	19.7	33		
WB3N05	33.6	33		
WB3N05	39.5	33.9		
WB3N06	31.8	16.6		
WB3N06	39.2	17.4		
WB3N07	24.1	30.2		
WB3N07	27	30.6		
WB4N01	27.7	14.8		
WB4N01	22.6	14.2		

Source: results of a simulation in the FlexSim software

6 Conclusion

Using a product input–output model adapted to the MFCA system can facilitate the process of implementation of modern matrix costing methods in industrial SMEs. The technological structure of the product input–output model makes it possible to determine the amount of unit costs for the individual components of the analyzed product with high precision. This advantage provides a good basis for short-term planning of production costs. In this case, the chosen planning strategy should reflect the requirements of the present day, which include, in particular, the need to reduce energy consumption and waste volume. Another benefit of product input–output models is their seamless integration with CAD software in order to obtain initial data for the construction of the first and second quadrants. This fact makes the product input–output model an effective tool that can be used not only by economists, but also by mechanical engineers. Determining the rejected product coefficient using equations derived from a product input–output model can facilitate the process of aggregation with an organizational–technological input–output model adapted to the MFCA system. Such an aggregated input–output model would, in the future, be a very powerful tool in the field of matrix costing for the requirements of the MFCA system. The MFCA system only allows the flow of the production material to be monitored as a whole, and not as a flow of components produced from given material. Such aggregation would lead to an overall refinement of matrix costing within the context of the MFCA system.

Future research will be directly focused on the integration of the designed product input–output model with the organizational–technological input–output model adapted to the MFCA system. As part of effective integration, great emphasis will be placed on the use of relevant software designed for the simulation of production processes as an important tool for testing the proposed methodologies based on matrix costing and Input–Output Analysis. Among the other goals of the research, the authors include the design of an organizational–technological input–output model that would be able to separate variable and fixed costs. Such a structural model could represent a very powerful tool suitable for the needs of variable costing.

Acknowledgments This work had been supported by the ČVUT Student Grant Competition No. SGS23/165/OHK2/3T/12.

References

1. V. Albino, S. Kühtz, Enterprise input-output model for local sustainable development – The case of a tiles manufacturer in Italy. Resour. Conserv. Recycl. **41**(3), 165–176 (2004). Available at https://doi.org/10.1016/j.resconrec.2003.09.006
2. S. Dierkes, D. Siepelmeyer, Production and cost theory-based material flow cost accounting. J. Clean. Prod. **235**, 483–492 (2019). Available at https://doi.org/10.1016/j.jclepro.2019.06.212
3. Z. Drábková, M. Pech, Comparison of creative accounting risks in small enterprises: The different branches perspective. E&M Econ. Manag. **25**(1), 113–129 (2022). https://doi.org/10.15240/tul/001/2022-1-007
4. J. Dubihlela, R. Rosebud, Employee training, managerial commitment and the implementation of activity based costing; impact on performance of SMEs. Int. Bus. Econ. Res. J. **13**(1), 27–38 (2014). Available at https://doi.org/10.19030/iber.v13i1.8354
5. A. Foroughi, M. Kocakulah, A. Stott, L. Manyoky, Activity-based costing: Helping small and medium-sized firms achieve a competitive edge in the global marketplace. Res. Econ. Manag. **2**(5), 150–171 (2017). Available at https://doi.org/10.22158/rem.v2n5p150
6. K. Janovská, I. Vozňáková, L. Švajdová, The verification of applicability of economical mathematics methods of structural analyses as a tool for optimising economic proceedings of metallurgical enterprise, in *METAL 2010*, (Tanger, Ostrava, 2010), p. 5. ISBN 978-80-87294-15-4
7. H.D. Kurz, N. Salvadori, 'Classical' roots of input-output analysis: A short account of its long prehistory. Econ. Syst. Res. **12**(2), 153–179 (2000)
8. W. Leontief, *Input-Output-Economics*, 2nd edn. (Oxford University Press, New York, 1986)

9. M. Machka, T. Beran, Using economical-mathematics method of input-output analysis as a tool for cost calculation of steel construction production, in *6th EAI International Conference on Management of Manufacturing Systems. EAI/Springer Innovations in Communication and Computing*, ed. by L. Knapčíková, D. Peraković, (Springer, Cham, Poland, 2022), pp. 257–272, Oct 6 (2021). Available at https://doi.org/10.1007/978-3-030-96314-9_17
10. A. Malik, M. Egan, M. Plessis, M. Lenzen, Managing sustainability using financial accounting data: The value of input-output analysis. J. Clean. Prod. **293** (2021). Available at https://doi.org/10.1016/j.jclepro.2021.126128
11. R.E. Miller, P.D. Blair, *Input-Output Analysis: Foundations and Extensions*, 2nd edn. (Cambridge University Press, New York, 2009)
12. P. Ruane, P. Walsh, J. Cosgrove, Validation of a digital simulation model for maintenance in a high-volume automated manufacturing facility. IFAC-Pap. **55**(19), 127–132 (2022). Available at https://doi.org/10.1016/j.ifacol.2022.09.195
13. M.P. Sadar, K.G. Rajmore, M.K. Rodge, K. Kumar, Digital manufacturing approach for process simulation and layout optimization. Mater. Today: Proc. **74**(4), 642–649 (2023). Available at https://doi.org/10.1016/j.matpr.2022.10.003
14. J. Slavíčková, O. Slavíček, Classification of company life cycle stages: An automotive industry in the Czech Republic. E&M Econ. Manag. **25**(4), 85–99 (2022). https://doi.org/10.15240/tul/001/2022-4-006
15. L.C. Stillwell, R.C.A. Minnitt, Input-output analysis: Its potential application to the mining industry. J. South. Afr. Inst. Min. Metall. **100**(7), 455–460 (2000)
16. R.R. Tan, K.B. Aviso, M.A.B. Promentilla, K.D.S. Yu, J.R. Santos, *Input-Output Models for Sustainable Industrial Systems: Implementation Using LINGO* (Springer, Singapore, 2018). Available at https://doi.org/10.1007/978-981-13-1873-3

Sustainability and Circular Economy Through Application and Processing of Recycled Materials into Additive Manufacturing

Rebeka Tauberová, Lucia Knapčíková, Daniel Strametz, and Muaaz Abdul Hadi

1 Introduction

Nowadays the concept of sustainability is an essential part of every field. And with the advancement of modern technologies, it is possible to ensure the basic pillars of sustainability. Transformation from linear model economy to circular model of economy is a process that every human being is a part of. From this point of view it is important to implement recycled materials into manufacturing processes such as additive manufacturing. High consumption of non-renewable energy sources and materials is a disastrous assumption for the future. Basic principle of linear economy is to produce goods, subsequent distribution, and consumption, and finally to produce waste. Waste that comes up from the valuable raw materials ends up in landfills or in incinerators. Today and in the future, it is important to stick with the useful idea that it is necessary to prevent the generation of waste in general. This is possible through recycling the materials. Implementation of recycled materials into various manufacturing processes can be challenging and, at the same time, a great opportunity. To prevent the accumulation of waste is a very important step but also very difficult to fulfill, mostly because of the complexity of whole recycling process, which is not well established in most countries. Excessive accumulation of waste in landfills can lead to several negative impacts on our ecological system and

R. Tauberová (✉) · L. Knapčíková · D. Strametz · M. A. Hadi
Faculty of Manufacturing Technologies with a Seat in Presov, Department of Industrial Engineering and Informatics, Technical University of Košice, Prešov, Slovakia

TU Graz, Faculty of Mechanical Engineering and Economic Sciences, Institute of Production Engineering, Graz, Austria

Pro2Future, GmbH, Graz, Austria
e-mail: rebeka.tauberova@tuke.sk; lucia.knapcikova@tuke.sk; daniel.strametz@tugraz.at; muaaz.abdulhadi@pro2future.at

L. Knapčíková, D. Peraković (eds.), *8th EAI International Conference on Management of Manufacturing Systems*, EAI/Springer Innovations in Communication and Computing, https://doi.org/10.1007/978-3-031-53161-3_2

environment. This paper focuses on the implementation of various plastic recycled materials into additive manufacturing processes and points out the importance of recycling in general [1–3].

1.1 State of the Art

In general, it is possible to admit that the amount of waste is rising enormously. Just because of that is it necessary to ensure that the largest possible part of the waste is somehow recycled and implemented as a secondary material in specific manufacturing processes. This paper focuses primarily on implementation of plastic recycled materials in additive manufacturing processes. In order to ensure the whole recycling process of plastic materials, there were established laws regarding the mandatory recycling of selected materials. Law no. 79/2015 Coll. Act on waste and on amendments and additions to certain laws regulates basic program documents in waste management, precautions to prevent generation of waste, rights and responsibilities of legal entities in waste management, extended producer responsibility, management of selected types of waste, municipal waste management, transboundary movement of waste, information system of waste management, and responsibility for breach of obligations arising from the aforementioned law. Based on aforementioned law, waste can be considered as a movable thing or substance. Recycled materials that will be mentioned in this paper were considered as a waste before they were recycled. Nowadays we distinguish two basic categories of waste management in general. First category is waste recovery, which includes, for instance, energy recovery, reuse, and recycling process. Second important category is waste disposal, this category includes landfill without subsequent recycling and waste incineration. One could state that from among the listed methods, recycling is the best one, because it is possible to breathe new life into the used material. Just because of the possibility to recycle used material and ensuring the careful reuse of specific materials, it is necessary to look for innovative ways to recover waste in general and especially to find interesting areas of use of the given recycled materials. The goal of presented study will be application of specific recycled materials for additive manufacturing. In the following section, Work Methodology discusses the novelty of proposed research [2, 4, 5].

1.2 Additive Manufacturing

Technology of additive manufacturing is well known among all industries. Basically, it is the process of creating physical object by layering and application of the specific material from 3D model. Additive manufacturing – AD can be defined as an antonym of traditional manufacturing, for instance, turning or milling. But the main advantage of AD is in global efficiency. It is possible to admit that

AD is more sustainable technology of producing some printed samples, mainly because of minimal production of waste during the manufacturing process of 3D printing. Therefore, it is possible to reduce the material which is entering the whole production process. And finally, it is desirable to enforce recycled materials into AD process, which is a great possibility. In general, there are several methods of AD; for instance, the most popular of all are FDM (Fused Deposition Modeling), SLA (Stereo Lithography), and SLS (Selective Laser Sintering). FDM method is most often used mainly because of its popularity and simple operation. In this most used method are thermoplastic materials melted and extruded through a nozzle layer by layer to build the specific 3D object. Figure 1 shows the FDM method.

In the upper part of Fig. 1, it is possible to see filament. Filament is a key component in AD. It is a long, thin, and flexible material, which is fed into a 3D printer and then melted, extruded through a nozzle to form the required object. Specific chosen type of filament can impact the final quality of printed samples. The most frequently used materials are plastics such as PLA and ABS. As well as are used most specific materials like nylon, PETG, or carbon fiber. However, other composite materials are also beginning to be used [6–9], which are often recycled

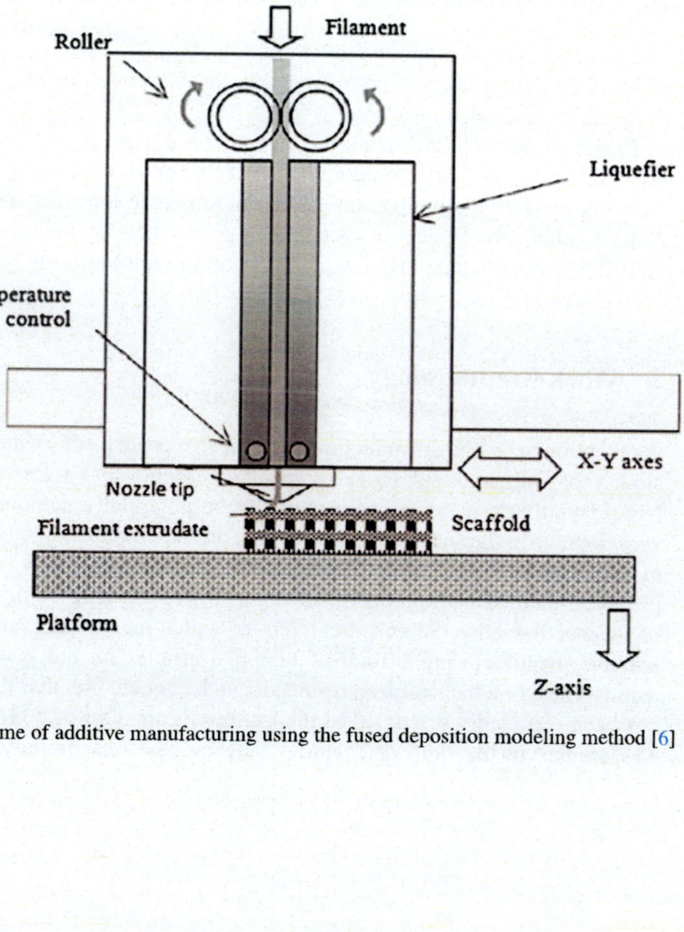

Fig. 1 Scheme of additive manufacturing using the fused deposition modeling method [6]

Fig. 2 Filament placed on 3D printer Creality CR – 5 Pro High – Temp [10]

or enriched with some reinforcing material that improves the mechanical properties of the pure material.

Figure 2 shows filament placed behind on 3D printer Creality CR – 5 PRO High Temperature. The main dominance of this printer is a fully enclosed chamber which ensures stable temperature. Maximum nozzle temperature is up to 300 °C. Among other advantages of specific 3D printer can be included large print size 300*225*380 mm, auto leveling, and support of cloud printing [10].

2 Work Methodology

As mentioned in the Introduction section, the concept of sustainability and possibility of utilization of recycled materials seems to be reasonable to analyze. Nowadays this area of research seems to be potentially demountable and will be very popular in the future. The aim of this presented paper is to connect technology of Additive manufacturing together with utilization of recycled materials. The proposed methodical approach has begun with analyzing of input data from Web of Science database. Searched entered keywords have been "recycled materials in additive manufacturing." Total of searched articles on the given issue were 834 results. Based on the analyzed results, it can be concluded, that this area of research has been repeatedly discussed in the last few years. Out of 834 results were found 348 articles in the field of Material Science Multidisciplinary; the second most

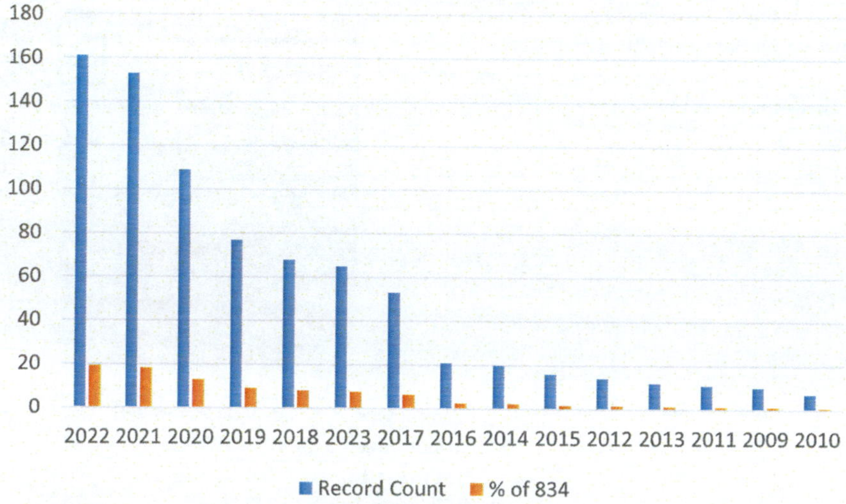

Fig. 3 Number of published articles in the specific area of research in the last decade [11]

frequently mentioned field of all results was Engineering Manufacturing with 141 articles. And 96 articles were found in the field of Environmental Sciences [11, 12]. The popularity of chosen topic confirms frequent occurrence in the last decade, as can be seen in Fig. 3.

According to the chart above, it can be concluded that the number of published articles in the area of proposed research has risen rapidly in the last 3 years. During 2022 alone, 161 articles were published from the total of 834. Nevertheless, in 2021, altogether 153 articles were published. Based on the data shown above, it can be assumed that utilization of recycled materials in additive manufacturing will be more circulating and forthcoming in general [13].

3 Recycled Materials in Additive Manufacturing

Nowadays, polymeric materials are an inseparable component of everyday life for every human in this world. By the 2050, it is estimated that plastic production will be around 1.1 billion tons [14]. Plastics are ubiquitous in modern culture mostly because of their excellent properties such as durability, lightweight, affordability, and it is wide enforcement in different industries, for instance in automotive, food industry and packaging, fashion industry and medicine [15, 16]. One could state that we are living in a plastic world because plastics and polymeric materials are everywhere. Twenty-five million tons of plastic waste are thrown away, while only 30% of this amount is recycled. Moreover, the global demand for plastic is growing

Fig. 4 Plastic waste management [17]

Table 1 Recycled filament for 3D printing

Filament type	Nozzle temperature	Bed temperature
Nonoilen (recycled)	175–195 °C	50 °C
PLA (recycled)	200 °C	60 °C
PETG (recycled)	245 °C	80 °C
TPU (recycled)	210 °C	85 °C
PVB (recycled)	200–225 °C	65–75 °C

year by year and the subsequent generation of plastic waste is expected to rise in direct proportion to global demand. So, it is possible to admit, that plastic waste management is key to solving the environmental and ecological issues. Figure 4 represents the plastic waste management [18–21].

To enhance the circularity of plastic materials, it is necessary to improve the recycling rates in every country and to ensure appropriate technology for the recycling process. By improving the recycling rates, it is possible to produce more recycled materials and therefore enforce these recycled materials into manufacturing processes [21].

The overall aim of submitted study is applied to utilization of plastic recycled materials into additive manufacturing. Implementation of recycled materials into additive manufacturing can bring a whole new perspective and other opportunities. As it was mentioned, the most commonly used material (filament) in 3D printing is PLA (polylactic acid). PLA is thermoplastic polyester, which is fully biodegradable.

Table 1 represents most used filaments, which are made from recycled materials and can be used for additive manufacturing process to create specific physical object [22].

Nonoilen is a material which is made from mixture of polylactic acid (PLA) and polyhydroxybutyrate (PHB). It is a fully biodegradable type of filament, which can be repeatedly recycled without losing its original mechanical properties. Nonoilen filament has similar properties as PLA filament and is easy to print from Nonoilen different physical objects. The content of PHB improves the toughness, temperature resistance, and there is no need for annealing. Decomposition of Nonoilen in compost is three times faster than PLA.

Fig. 5 Filament from recycled polyvinyl butyral

Recycled PLA is biodegradable polymer, this bio-sourced homopolymer can be obtained from corn starch. Polylactic acid is a natural alternative to polyethylene. Its main advantage is that recycled PLA is easy to print from it and the price is affordable. It can be used in industrial applications, notably packaging and in domestic applications [23].

Recycled PETG is made from high quality medical waste. This material is durable and, as well as other recycled materials, is easy to use with exceptional visual and mechanical performance.

Recycled TPU belongs to the polyurethane family of thermoplastics. It is translucent filament with good elasticity. It is used to print resistant and flexible physical objects. This type of recycled filament is especially used in automotive industry, in electronics, and consumer goods sector [17].

Recycled PVB is polymeric material which is transparent. Advantage of PVB is its mechanical properties. It has high tensile strength, impact resistance, and high adhesion [5, 24] (Fig. 5).

3.1 *Utilization of Extruder in Material Recycling Process*

Another alternative approach on how to recycle used filament is through utilization of the extrusion method. Filament extruder is an instrument, which is designed to create a feedstock for 3D printer from raw plastic pellets. Extruder is a device, which makes it possible to melt granulated material pellets and push the molten plastic through a nozzle, creating the filament. Produced filament can be used as an entering material for additive manufacturing processes. Labeled Fig. 6 represents the process of utilization of extruder in additive manufacturing.

Figure 7 labels the key features of the extruder Protocycler+. The mentioned device consists of a mechanical grinder and a grinder crank which can shred and

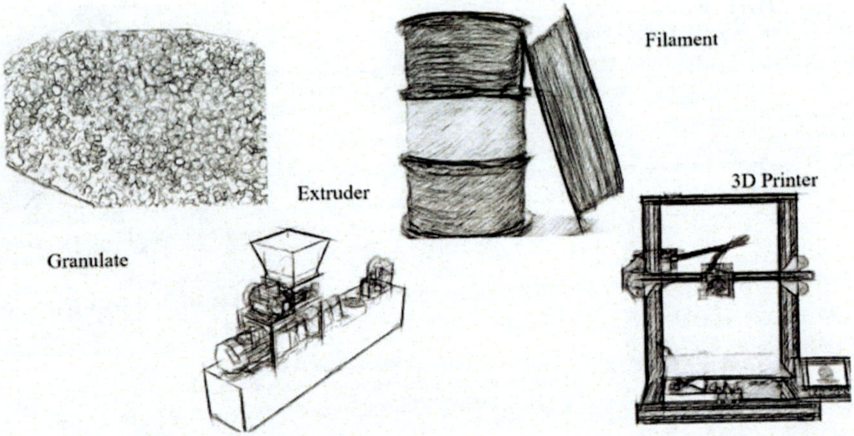

Fig. 6 Block scheme of extrusion process in additive manufacturing

Fig. 7 Front view and key features of extruder Protocycler+ [25]

unneeded 3D prints, 3D printing waste, and granules that can be used for new filament production. Another important part of the extruder is the extruder hooper, which is used as entry station for material that is entering the filament production process. Front view of extruder consists of control area, in which is placed extruder nozzle, diameter sensor, cooling fan, light guides, puller, and spreader.

In order to achieve the required quality of filament production, it is necessary to follow the established principles. In general, it is possible to admit, that the extrusion process has great economic importance, being the most used technique for polymer processing nowadays. The advantage of the extruder is the possibility to produce own filament from several types of polymers, even recycled. Figure 8 shows 3D printed waste and granulate, which can be used as entering material for the extrusion process [7, 26–28].

Fig. 8 3D printed waste and granulate for extrusion process

4 Results and Discussion

Implementation of recycled materials in additive manufacturing process is a prevalent and ongoing topic and will be in future research. Nevertheless, it is important to analyze the basic properties of specific recycled materials and to compare them with the virgin one. Considering the possibility that with the repeated use of recycled material, the basic properties may degrade eventually. From another point of view, the number of repeated cycles of the recycling process itself can be significant. The assumption that after the second or third cycle of recycling, the material loses the primordial properties is very debatable. Regularly it depends on the specific material that is being recycled and, moreover, recycled material can be enriched by some reinforcing materials. There are some obstacles, which can occur when recycled material is used as filament in additive manufacturing [27, 29, 30]:

- Irregular diameter – In general, the required filament diameter should be 1.75 mm, but while producing the recycled filament, a slight deviation may occur.
- Impurity of filament – During the recycling process, it is possible that some pollutant can get mixed into the filament; impurities in filament can cause random thickening.
- The possibility of breakage – During the recycling process, mixing of different materials occurs; entering materials have disparate quality and for this reason, local tension in filament may occur.
- Insecure color – Due to the mixing of several materials, the color changeability usually arises.
- Ovality – Fluidity of the molten material is disparate and because of the mixing of entering materials, this can result in the filament not being rounded but oval.

Figure 9 represents the example of irregular diameter. In some cases, a deviation of 0.5 mm is allowed.

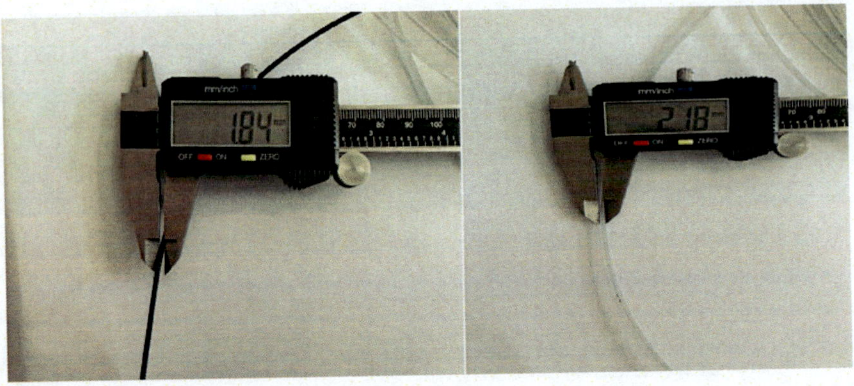

Fig. 9 Irregular diameter in filament

5 Conclusion

The intention of submitted study is to analyze the possibilities of implementing recycled materials into additive manufacturing processes. Considering, that mentioned technology is now ongoing and will be prevalent in the future together with the utilization of recycled materials is reasonable to discuss. Based on the literature, one can state that the relevance of this area of research has expanded in recent years. In the beginning, this paper points out the importance of circular economy, sustainable production, and circularity of old and used materials. The relation between additive manufacturing and utilization of recycled materials is significant because it enables to produce and print samples with zero waste and helps to preserve the value of virgin materials for longer periods. Fulfillment of recycled materials in additive manufacturing ensures saving of non-renewable resources. The primary focus of this paper is directed toward existing recycled materials that are being used in additive manufacturing as entering materials – filament for production process. Further work discusses the other capabilities for recycling the 3D printing waste or failed printed samples. The future direction of work will focus more on examining the properties of recycled materials in a more detailed way [1, 24].

Acknowledgments This work was supported by the project VEGA 1/0268/22, granted by the Ministry of Education, Science, Research and Sport of the Slovak Republic.

References

1. Inštitút Cirkulárnej Ekonomiky, Cirkulárna Ekonomika. https://www.incien.sk/cirkularna-ekonomika/. Last accessed 2023/06/07
2. T.R. Walker, L. Fequet, Current trends of unsustainable plastic production and micro (nano) plastic pollution. TrAC Trends Anal. Chem. **160**, 116984 (2023)

3. E.N. Kalali, S. Lotfian, M.E. Shabestari, S. Khayatzadeh, C. Zhao, H.Y. Nezhad, A critical review of the current progress of plastic waste recycling technology in structural materials. Curr. Opin. Green Sustain. Chem **40**, 100763 (2023)
4. N.P. Kim, 3D-printed conductive carbon-infused thermoplastic polyurethane. Polymers **12**(6), 1224 (2020)
5. I.S. Lase, D. Tonini, D. Caro, P.F. Albizzati, J. Cristóbal, M. Roosen, et al., How much can chemical recycling contribute to plastic waste recycling in Europe? An assessment using material flow analysis modeling. Resour. Conserv. Recycl. **192**, 106916 (2023)
6. S. Singh, S. Ramakrishna, R. Singh, Material issues in additive manufacturing: A review. J. Manuf. Process. **25**, 185–200 (2017)
7. J. Husár, L. Knapčíková, J. Kaščak, Optimization of the production process and defect identification in the production of face shields using fused deposition modeling technology, in *6th EAI International Conference on Management of Manufacturing Systems*, (Springer International Publishing, Cham, 2022, June), pp. 141–153
8. T.D. Ngo, A. Kashani, G. Imbalzano, K.T. Nguyen, D. Hui, Additive manufacturing (3D printing): A review of materials, methods, applications, and challenges. Compos. Part B **143**, 172–196 (2018)
9. R.B. Kristiawan, F. Imaduddin, D. Ariawan, Z. Arifin, A review on the fused deposition modeling (FDM) 3D printing: Filament processing, materials, and printing parameters. Open Engineering **11**(1), 639–649 (2021)
10. Creality, CR-5 Pro High-temp Version. https://www.creality.com/products/creality-cr-5-pro-h-3d-printer. Last accessed 2023/06/07
11. Y.F. Lv, W. Thomas, R. Chalk, S. Singamneni, Flame retardant polymeric materials for additive manufacturing. Mater. Today Proc **33**, 5720–5724 (2020)
12. O.S. Saran, A.P. Reddy, L. Chaturya, M.P. Kumar, 3D printing of composite materials: A short review. Mater. Today Proc **64**, 615 (2022)
13. WebOfScience, Analyze results. https://www.webofscience.com/wos/woscc/analyze-results/124d5788-b6fc-4ec2-853a-a5d1fa363a20-8fe64219. Last accessed 2023/06/07
14. N. Van de Werken, H. Tekinalp, P. Khanbolouki, S. Ozcan, A. Williams, M. Tehrani, Additively manufactured carbon fiber-reinforced composites: State of the art and perspective. Addit. Manuf. **31**, 100962 (2020)
15. A. Al Rashid, S.A. Khan, S.G. Al-Ghamdi, M. Koc, Additive manufacturing of polymer nanocomposites: Needs and challenges in materials, processes, and applications. J. Mater. Res. Technol. **14**, 910–941 (2021)
16. Turtle, Homepage, https://www.turtle.sk/sk/961/16/recyklovany-pla-cierny-1-kg. Last accessed 2023/06/06
17. M. Bustos Seibert, G.A. Mazzei Capote, M. Gruber, W. Volk, T.A. Osswald, Manufacturing of a PET filament from recycled material for material extrusion (MEX). Recycling **7**(5), 69 (2022)
18. G. Barnabas, A. Geethan, A. Vasantha, Analysis of material properties of recycled plastic and rubber components. Fresenius Environ. Bull. **30**, 1353–1362 (2021)
19. P. Parandoush, D. Lin, A review on additive manufacturing of polymer-fiber composites. Compos. Struct. **182**, 36–53 (2017)
20. E. Pakdel, S. Kashi, R. Varley, X. Wang, Recent progress in recycling carbon fibre reinforced composites and dry carbon fibre wastes. Resour. Conserv. Recycl. **166**, 105340 (2021)
21. E.D. Okoffo, E. Donner, S.P. McGrath, B.J. Tscharke, J.W. O'Brien, S. O'Brien, et al., Plastics in biosolids from 1950 to 2016: A function of global plastic production and consumption. Water Res. **201**, 117367 (2021)
22. H. Wu, Enhancements of sustainable plastics manufacturing through the proposed technologies of materials recycling and collection. Sustain. Mater. Technol. **31**, e00376 (2022)
23. R. Torres, Z. Cheng, R. Ramalingame, O. Kanoun, Electrical characterization of elongation sensors based on SBS-CTPU filaments, in *2018 15th International Multi-Conference on Systems, Signals & Devices (SSD)*, (IEEE, 2018, March), pp. 1212–1215

24. E.S. Lakatos, G. Yong, A. Szilagyi, D.S. Clinci, L. Georgescu, C. Iticescu, L.I. Cioca, Conceptualizing core aspects on circular economy in cities. Sustain. For. **13**(14), 7549 (2021)
25. ReDeTec, ProtoCycler +.: Extrusion Theory Guide. https://redetec.com/pages/guides-and-manuals. Last accessed 2023/06/06
26. F.A. Cruz Sanchez, H. Boudaoud, S. Hoppe, M. Camargo, Polymer recycling in an open-source additive manufacturing context: Mechanical issues. Addit. Manuf. **17**, 87–105 (2017)
27. P. Byrley, M.A.G. Wallace, W.K. Boyes, K. Rogers, Particle and volatile organic compound emissions from a 3D printer filament extruder. Sci. Total Environ. **736**, 139604 (2020)
28. S.N. Priya, S.N. Kumar, S.P. Kumar, K.K. Pradeep, Design and fabrication of filament extruder with spooler. Mater. Today Proc **81**, 221 (2021)
29. B. Almeshari, H. Junaedi, M. Baig, A. Almajid, Development of 3D printing short carbon fiber reinforced polypropylene composite filaments. J. Mater. Res. Technol. **24**, 16–26 (2023)
30. A.K. Gupta, M. Taufik, Effect of process variables on performances measured in filament and pellet-based extrusion process. Mater. Today Proc **47**, 5177–5184 (2021)

Contemporary Approach of Additive Manufacturing for Sustainable Products in the Era of Circular Economy

Lucia Knapčíková, Annamária Behúnová, Rebeka Tauberová, Jakub Kaščak, and Jozef Husár

1 The State of the Art

Sustainable production supports environmentally impacting activities in order to minimize pollution and supports emerging sustainable innovations [1]. Sustainable production is also known as ecological production, which aims to implement activities that influence the design of processes and product concepts [1, 2]. For example, the automotive industry is also highly competitive. Industrial enterprises actively take measures to increase productivity and minimize waste [5, 6]. They reduce product processing time and minimize the number of unnecessary products [2]. The main benefit is the improvement of equipment and process performance. It is more reliable and can be automated, which also provides protection against various risks [3]. Sustainable production can also be characterized as a method that offers companies opportunities to improve the image of the industry, increase customer loyalty, and achieve a higher position in the market [5, 6]. With increasing emphasis on sustainable development worldwide, many customers are looking for

L. Knapčíková (✉) · R. Tauberová · J. Husár
The Technical University of Košice, Faculty of Manufacturing Technologies with a seat in Prešov, Department of Industrial Engineering and Informatics, Prešov, Slovakia
e-mail: lucia.knapcikova@tuke.sk; rebeka.tauberova@tuke.sk; jozef.husar@tuke.sk

A. Behúnová
Technical University of Kosice, Faculty of Mining, Ecology, Process Control and Geotechnologies, Institute of Earth Resources, Kosice, Slovakia
e-mail: annamaria.behunova@tuke.sk

J. Kaščak
The Technical University of Košice, Faculty of Manufacturing Technologies with a seat in Prešov, Department of Computer Aided Manufacturing Technology, Prešov, Slovakia
e-mail: jakub.kascak@tuke.sk

L. Knapčíková, D. Peraković (eds.), *8th EAI International Conference on Management of Manufacturing Systems*, EAI/Springer Innovations in Communication and Computing, https://doi.org/10.1007/978-3-031-53161-3_3

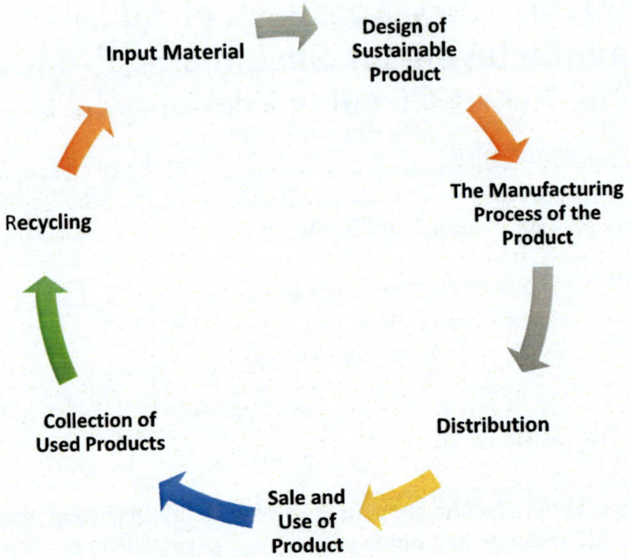

Fig. 1 Life cycle of the product in the context of circular economy. (Authors own processing)

specific companies, which are sustainable and ensure that the company achieves sustainability goals [5]. Sustainable production is necessary to reduce the amount of waste, raw materials, electricity 1.2 g, packaging, and other resources [1, 4, 7]. This is a new approach to implementing a waste reduction method. This helps reduce the amount of waste by recycling materials (Fig. 1).

Sustainable production supports environmentally impacting activities in order to minimize pollution and supports emerging sustainable innovations [3, 4]. Sustainable production is also known as ecological production, which aims to implement activities that influence the design of processes and product concepts [8]. For example, it is also highly competitive in the automotive industry [5]. Industrial enterprises actively take measures to increase productivity and minimize waste. They reduce product processing time and minimize the number of unnecessary products. The main benefit is the improvement of equipment and process performance [3, 5]. It is more reliable and can be automated, which also provides protection against various risks [9]. Industry 4.0 is a vast, constantly evolving industrial revolution, the most representative technology of which is additive printing technology [5].

1.1 Advantages of Additive Technology in the Industry 4.0 Concept

Additive manufacturing compared to conventional manufacturing technologies such as co-turning, where the material is removed with a knife, the product is manufactured by adding material – from the English "adding" – to add. As a result, additive technology does not produce or produces minimal waste. The advantages are in the area [5, 10]:

- Reduction in production times and costs.

Thanks to additive technologies, the creation of a prototype is faster and cheaper compared to milling, where demanding configuration and material preparation is required.

- It supports the digitization of the company.

Additive technology requires constant contact between devices. This is only possible thanks to the digitization of production equipment.

- It combines the assembly process into one part.

Another benefit of additive manufacturing is the simplification of assembly and completion of products. Conventional components are more complex and require more manufacturing steps. This increases the cost of material and labour and requires more time. Additive technology enables the assembly to be printed in one part.

Compared to conventional production processes, additive manufacturing has the following four basic values [4, 5]:

The ability to create almost any shape allows designers to create machine parts/spare parts that perform better and cost less.

- It is possible to create different shapes without the need for moulds, which opens up possibilities for a customized shape from the designer's request.
- Eliminating time-consuming production operations, shortening the time to launch the product on the market.
- Additive technology can simplify the maintenance of stands, thereby reducing the need for spare parts stocks.

The field of application of additive technology is mainly in the medical industry [4].

Since the medical industry is developing the fastest, it uses additive technology to help patients. Medical device manufacturers use a wide range of high-strength and bio-compatible additive printing materials, from flexible to rigid and from semi-transparent to opaque. Applications that revolutionize the medical industry are pre-operative models from CT images, orthopaedic implants, and dental prostheses [5, 11].

***Energy Industry* [4]**
The development of critical elements that can withstand extreme conditions is important for the energy industry. For the needs in the field, additive technology was helpful in the production of efficient, lightweight components and environmentally friendly materials. Key applications that have emerged in the oil, gas, and energy sectors include rotors, turbine nozzles, and downhole tool components.

***Aerospace Industry* [4]**
The aerospace industry is considered the first sector to adopt and use additive technology in its practice. Some of the most demanding performance standards are used in this sector, which require parts to last in extreme conditions. Common applications include rocket engine components and custom aircraft interior components [12]. Sustainable production can also be characterized as a method that offers companies opportunities to improve the image of the industry, increase customer loyalty, and achieve a higher position in the market. With the growing emphasis on sustainable development around the world, many customers are looking for specific companies that are sustainable and ensure that the company achieves sustainability goals.

2 Work Methodology

When choosing an additive technology process for printing experimental samples, emphasis was placed on the smallest possible energy expenditure. Therefore, Fused Deposition Modelling or FDM, was chosen. It is one of the most used production processes for the production of prototypes [5, 6]. This process is based on heating the nozzle and extruding the filament onto the substrate in thin layers [13]. Due to the simplicity, reliability, and affordability of the FDM process, this additive manufacturing technology has become widely recognized and accepted by industry, academia, and consumers.

Advantages of using FDM portion [6, 14]:

- Reduced waste: Thanks to this technology and precise layer placement, the Slicer is designed to calculate the exact amount needed to print an experimental sample.
- Reduced emissions: FDM additive manufacturing enables printing even in smaller office spaces and is financially undemanding. Distributed additive manufacturing can reduce emissions from transport by making it possible to print the desired models on-site.

Disadvantages of using the FDM process [4, 6, 15]:

- Surface unevenness: during printing and layering, surface unevenness can occur caused by the slightest movement of the pad or higher air humidity.
- Need for calibration: the FDM additive printer needs to be calibrated before each print.

2.1 Definition of Input Materials

Different brands and types of PLA filament have different printing behaviour. The right PLA filament should be consistent and suitable for the printer being used [4].

Another parameter influencing the choice of the used filament is its properties. Some PLA filaments have special properties, such as increased strength, elasticity or resistance to UV radiation [7]. Choosing the right PLA filament with the desired properties for the test samples is important. Some cheap PLA filaments may contain dangerous ingredients such as heavy metals or phthalates. Therefore, it is necessary to emphasize the correct choice of material from a renowned manufacturer that guarantees the safety of its products [3].

Filament from Creality 1.75 mm PLA
This filament from Creality is made using premium raw materials, and thanks to its properties such as high strength and elasticity, it is ideal for the production of experimental samples (Fig. 2). The manufacturer guarantees long-term colour stability [4].

Advantages

- It is a gentle, non-toxic, biodegradable material derived from renewable sources such as corn.
- It has a lower tendency to deformation and cracking than ordinary PLA filament, which makes it a more stable material for larger and more complex shapes [16].
- It is relatively easy to print and has low demands on the printer, which makes it an ideal material for beginners.
- It is made from premium raw materials and has low toxicity, which makes it a safe material for use in 3D printers in a home environment [5, 6].

Disadvantages

- It has a lower heat resistance than some other materials, such as ABS, which means that it is not suitable for the production of objects that are exposed to higher temperatures as a powerful material, which is also reflected in its higher price [12].

Fig. 2 The filament Creality PLA 1.75 [4]

Fig. 3 The filament Creality PETG 1.75 [4]

- It is more sensitive to moisture and other storage factors, which can lead to its early deterioration and loss of quality [6, 17].

Filament from Creality 1.75 mm PETG
This filament from Creality is characterized by its maximum flexibility. The manufacturer repeatedly declares the filament's flexibility without cracking [4, 6]. The resulting product subjected to additive manufacturing is perfectly smooth on the surface (Fig. 3). Chemical vapours are not produced during printing [18].

Additive printer Ender CR-10 – An Ender CR-10 printer (Fig. 4) was used to create test samples for this work. The Ender CR-10 printer from Creality is one of the best options when choosing an inexpensive printer [18, 23].

It is a printer with a large printing range. This printer uses the Fused Deposition Modelling printing method. It is possible to print models with dimensions of 300 × 300 × 400 mm [4, 8, 19]. The frame of the printer is made of aluminium construction, which ensures its stability. Suitable materials for use in this printer are PLA, ABS, TPU, and PETG with a filament diameter of 1.75 mm [8, 20].

3 Results and Discussion

The investigated samples' shape and size were considered for the size and shape of the seeds that were inserted into the spaces between the fillings (into the hollow holes) in the experimental sample during printing. The samples were designed in CAD software Autodesk Fusion 360 and imported in "stl" format into print preparation software – this software is called Slicer [4, 9, 21].

Properties of test samples:

- Dimensions of the experimental sample: 30 × 50 × 20mm
- Wall and base thickness: 2 mm

Fig. 4 The filament Creality PETG 1.75. (Authors own processing)

- Used % of the filling (infill density) – 10%
- Used filling: rectilinear, grid, triangular, wiggle, fast honeycomb, full honeycomb (Fig. 5) For experimental samples' research and comparison, these samples were divided according to the material and type of internal filling [4, 22, 23]. For comparison, six models made of PLA material and six models of PETG material were created, while each sample was printed with a different type of filling: rectilinear, grid, triangular, wiggle, fast honeycomb, and full honeycomb. When printing experimental ones, it was necessary to insert plant seeds into the hollow fillings of the model [9, 10, 24]. Due to limited space, small ornamental pumpkin seeds were placed in the holes. During the printing process, after printing half of the material, seeds were placed in the holes with tweezers. It was hypothesized (due to the bio-degradable properties of PLA plastic) that the experimental model could show a significant decomposition stage that would lead to the germination of the seed stored in the hollow holes [11, 25].

In the AutoDesk Fusion 360 software, a flower pot model with a hollow bottom in the shape of an arch was designed. The material required for the production of this planter will be biodegradable plastic PLA [4, 11, 26]. The hollow double bottom contains hollow fillings with a Honeycomb Full structure, in which seeds can be stored during printing (Fig. 6).

This product can be used immediately after printing. Any plant of the consumer's choice can be planted inside the pot, and the pot must be planted in the soil in such a way that the upper edge of the pot merges with the soil [13, 27]. It was considered that if, after the gradual death of a plant planted in a pot, the pot remains in the soil,

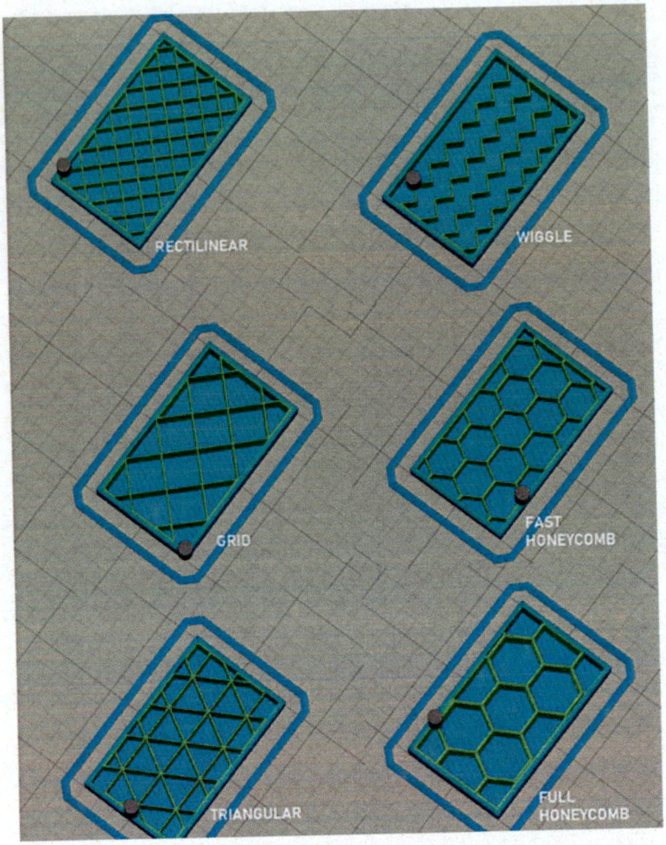

Fig. 5 Design and visualization of used filling [4]

a new plant will grow from the seeds after the decomposition of the double bottom [27].

Thanks to the wide availability of PLA filaments on the market, it is possible to produce a flower pot in various coloured versions and in size adapted to the given additive printer on which the product will be printed [14, 16].

Advantages of the Proposed Product

- It is a type of sustainable plastic product that does not need to be disposed of after use or damage.
- It is made of PLA plastic, which is made from corn starch and does not contain toxic substances that could seep from the soil into groundwater.
- No waste material is created during the production of the product [4, 16].

Fig. 6 Product design used software AutoDesk Fusion 360 [4]

4 Conclusion

The circular economy is considered to trigger a radical change in production systems with the ultimate goal of preserving the value of resources in the enterprise and eliminating waste. Moving towards a circular economy involves adopting cleaner production methods and relying on renewable energy sources and digital technologies. On the other hand, it involves the elimination of toxic materials, chemicals and focuses on designing durable and repairable products. While the circular economy is considered a prerequisite for a sustainable economy, none of its strategies is sustainable by default and needs to be monitored before, during, and after implementation to assess the impact of sustainability. Sustainable production is necessary to reduce the amount of waste, raw materials, electricity, paper, packaging, fuel, and other resources. This is a new approach to implementing a waste reduction method. This helps reduce waste by recycling materials – the paper aimed to justify using the additive method as part of the fourth industrial revolution. In companies, additive manufacturing has the potential to reduce the economic and environmental impact by reducing electricity consumption and material consumption.

Acknowledgements This work was supported by the project VEGA 1/0268/22-granted by the Ministry of Education, Science, Research and Sport of the Slovak Republic.

References

1. C.M.F. Barry, S.A. Orroth, Processing of thermoplastics, in *Modern Plastics Handbook*., ISBN 0-07-026714-6, ed. by C.H.A. Harper, (2000)
2. Circular economy: definition, importance and benefits. 2023. [online], [10.06.2023]. Available on https://www.europarl.europa.eu/news/en/headlines/economy/20151201STO05603/circular-economy-definition-importance-and-benefits. Cited 23.04.2023
3. J. Liu, A.T. Gaynor, S. Chen, Z. Kang, K. Suresh, A. Takezawa, L. Li, J. Kato, J. Tang, C.C.L. Wang, et al., Current and future trends in topology optimization for additive manufacturing. Struct. Multidiscip. Optim. **57**, 2457–2483 (2018)
4. Z. Dicka, Increasing product competitiveness using innovative production technology Master thesis. TU Košice (2023)
5. L. Wang, H. Zhou, N. Li, Carrier transport composites with suppressed glass-transition for stable planar perovskite solar cells. J. Mater. Chem. A **8**(28), 14106–14113 (2020)
6. G. Heath, T. Silvermann, M. Kempe, et al., Research and development priorities for silicon photovoltaic module recycling to support a circular economy. Nat. Energy **5**(7), 502–510 (2020)
7. T. Mandičák, P. Mesároš, M. Tkáč, Impact of management decisions based on managerial competencies and skills developed through BIM technology on performance of construction enterprises. Pollack Periodica **13**(3), 131–140 (2018)
8. M. Straka, A. Rosova, et al., Principles of computer simulation design for the needs of improvement of the raw materials combined transport system. Acta Montan. Slovaca **23**, 163–174 (2018)
9. S.A. Kaufmann, *The Origins of Order: Self-Organization and Selection in Evolution* (Oxford University Press, New York, 1993)
10. W.F. Hosford, *Mechanical Behavior of Materials* (Cambridge University Press, New York, 2010)
11. N. Dodds, A.G. Gibson, D. Dewhurst, J.M. Davies, Fire behavior of composite laminates. Compos. Part A. **31**, 689 (2000)
12. S. Fiebig, J. Sellschopp, H. Manz, T. Vietor, K. Axmann, A. Schumacher, Future challenges for topology optimization for the usage in automotive lightweight design technologies, in *Proceedings of the 11th World Congress on Structural and Multidisciplinary Optimization, Sydney, Australia, 7–12 June 2015*, vol. 142, (2015), pp. 1–8
13. R. Singh, J.P. Davim, *Additive Manufacturing: Applications and Innovations*, 1st edn. (CRC Press, Boca Raton, 2018), p. 280
14. N. Li, S. Huang, G. Zhang, R. Qin, W. Liu, H. Xiong, G. Shi, J. Blackburn, Progress in additive manufacturing on new materials: A review. J. Mater. Sci. Technol. **35**, 242–269 (2019)
15. J. Sinay, M. Balazikova, M. Dulebova, et al., Measurement of low-frequency noise during CNC machining and its assessment. Measurement **119**, 190–195 (2018)
16. A. Kanáliková, A. Behúnová, T. Mandičák, P. Mesároš, New technologies for sustainable development in infrastructure, logistics and construction industry. Acta Logistica **6**(4), 171–177 (2019)
17. M. Al-Salamah, Economic production quantity in batch manufacturing with imperfect quality, imperfect inspecyion and non-destructive acceptance sampling in a two-tier market. Comput. Ind. Eng. **93**, 275–285 (2016)
18. C.C. Lin, T.H. Wang, Build-to-order supply chain network design under supply and demand uncertainties. Transp. Res. B **45**, 1162–1176 (2011)
19. A. Smeringaiova, M. Wittner, Parametric modeling as efficient way of designing. MM Sci. J **12**, 2701–2705 (2018). https://doi.org/10.17973/MMSJ.2018_12_201873
20. R.T. Grenci, C.A. Watts, Maximizing customer value via mass customized e-customer services, in *Business Horizons*, (2007), pp. 123–132
21. Transparency market research [online], [10.06.2023]. Available on http://www.transparencymarketresearch.com/polyvinyl-butyral-films-sheets.html

22. M.D. Guenov, Complexity and cost effectiveness measures for systems design, in *Manufacturing Complexity Network Conference 2002 – Proceedings*, (Cambridge, 2002), pp. 455–466
23. A.M. Kaplan, M. Haenlein, Toward a parsimonious definition of traditional and electronic mass customization. J. Prod. Innov. Manag. **23**, 168–182 (2006)
24. F. Herrmann, The smart factory and its risks. Systems **6**, 38 (2018). https://doi.org/10.3390/systems6040038
25. I. Cvitić, D. Peraković, M. Periša, M. Botica, Novel approach for detection of IoT generated DDoS traffic. Wirel. Netw **1** (2019). https://doi.org/10.1007/s11276-019-02043-1
26. M. Kornaszewski, R. Pniewski, The impact of new telematics solutions on the safety of railway traffic on the example of modern simulators railway traffic control devices, in *Development of Transport by Telematics*, ed. by J. Mikulski, (Springer, Cham, 2019), pp. 32–43
27. I.I. Iatcheva, A.D. Andreev, R.D. Stancheva, I.T. Lilyanova, Electromagnetic flow meter field distribution maximizing device sensitivity. Mater. Sci. Forum **856**, 157–162 (2016)

Cost Operation Optimization with Binary Integer Linear Programming in a Cosmetic Company

Fredy Altamirano and Diego Vallejo-Huanga

1 Introduction

Cosmetic products are widely spread in different markets and are part of the daily life of a large part of the world population. The purpose of dermatological and cosmetic products is to help improve the skin's appearance, health, and odor. In addition, these products are used by consumers of all ages, and there are a large number of products that, depending on the area of application on the body, have a therapeutic or hygienic effect, e.g., deodorants, skin or hair care products, makeup, sunscreen, etc. [13].

The term cosmeceutical refers to a category of products with both pharmaceutical and cosmetic properties; its active ingredients generate a pharmacological action, which triggers a beneficial physiological effect in the area where it is applied [9]. Generally not subject to rigorous testing and regulation, as is required for pharmaceutical products. Lately, cosmeceutical products have been popular due to various studies and products on the market, offering solutions that provide benefits not only for beautification and skin care but also sanitizing, antiseptic, anti-inflammatory,

This work was supported by the IDEIAGEOCA Research Group of the Universidad Politécnica Salesiana in Quito, Ecuador.

F. Altamirano
Universidad Politécnica Salesiana, Production and Industrial Operations, Quito, Ecuador
e-mail: faltamiranoc@est.ups.edu.ec

D. Vallejo-Huanga (✉)
Universidad Politécnica Salesiana, Production and Industrial Operations, Quito, Ecuador

Universidad Politécnica Salesiana, IDEIAGEOCA Research Group, Quito, Ecuador
e-mail: dvallejoh@ups.edu.ec

L. Knapčíková, D. Peraković (eds.), *8th EAI International Conference on Management of Manufacturing Systems*, EAI/Springer Innovations in Communication and Computing, https://doi.org/10.1007/978-3-031-53161-3_4

depigmenting, antioxidant effects, among others [17]. Usually, these products are over the counter, i.e., they do not require a prescription to be sold directly to the consumer. Free access to these products and the wide range of brands have influenced users' consumption habits since the use of recognized cosmetics is linked to social and economic status. Mafra et al. [11] state that an individual's self-esteem is closely related to consumption and social status. Purchasing this product type can achieve happiness for particular consumers [10].

Since the beginning of the nineteenth century, globalization has had a social and cultural influence on cosmetics consumers worldwide, mainly because companies translated social values into brands; as a result, social perceptions of beauty changed [8]. This influence in Ecuador is marked by the US market, which has introduced beauty standards and influenced consumer behavior for this type of product. The search for adaptation to imported beauty canons, social acceptance, and the growing concern for skin care have driven the growth of the cosmetics industry in Ecuador. Currently, this market generates annual profits that exceed 1.1 billion dollars a year, with a yearly increase of 10% [12].

Currently, the growth strategies of the dominant cosmetic companies worldwide are based on the launch of selected products for the target market, the sustained marketing of their leading brands, and the hiring of local labor to provide an experience much closer to the consumer, the main strategies. The major commercial expansion of these cosmetic brands is focused on regions such as South America, Eastern Europe, Asia, and especially China. These markets cover consumption greater than 10.4% worldwide [10].

Interest in beauty care has increased due to the growth of the middle class in Ecuador. For most of these consumers, personal care is not optional but a necessity. Although the acquisition of hygiene and beauty items is widespread, the demand for national cosmetics and products is low due to the lack of quality and control by local regulatory agencies [12]. Until 2014, revenue from the cosmetics sector represented 1.6% of the gross domestic product. For this reason, based on the policy of change of productive matrix, the Ecuadorian government promoted the country's internal production growth for these products intending to have a broader internal capital flow. This policy has fostered the creation of sources of employment in different provinces. To achieve this change, companies in the sector have invested heavily in plants, inputs, and personnel training, intending to meet the quality standards that these products require [12].

The demand for cosmetics continues to rise despite the economic downturn due to the global pandemic. According to the Association of National Manufacturers, Marketers, and Importers of Cosmetic Products (Procosméticos), local purchasing behaviors and international reports show that the demand for personal hygiene products has grown 60% [3]. In Ecuador, Small and Medium Enterprises (SMEs), family businesses, and catalog sales are the primary sources of growth for the industry. However, due to the insufficient quality and supply of the product, the market is still highly dependent on imports [12]. Due to the growth of the cosmetic sector in the country and the benefits it brings to the state and consumers, the need to efficiently manage their resources arises in companies. This has made the cosmetic industry and manufacturing adopt planning and scheduling as an essential technical resource to manage its resources and human and financial capital [4]. According to Heizer and Render [7], optimal use of labor, equipment, and space translates into properly managing company resources. This allows efficient financial management of the organization and can be achieved by implementing a planning and programming program.

On the other hand, linear programming is part of the mathematical programming models. It is necessary to formulate them to identify the variables involved, their restrictions, and one or more functions that allow for optimizing a process. The restrictions include the availability of resources, technical specifications, or other conditions [15]. Linear programming can be used in various areas of production and industrial operations. The elaboration of production optimization models supports strategic decision-making applicable to any organization, especially SMEs, which require a financial tool that efficiently allocates the resources [5].

Several investigations seek to optimize production planning, scheduling, and the supply chain. The work [14] refers to a mathematical model through linear programming and the theory of restrictions to achieve optimal manufacturing quantities at the lowest possible cost. The research by Villa and Taurino [21] generates a system for designing production programs, considering aspects such as the production sequence, the date the orders are released, and the raw materials and the delivery date. That must be fulfilled. Their study focuses and applicability were SMEs. On the other hand, Tan et al. [16], using a fuzzy linear programming model, managed to determine the optimal adjustments in production levels, identifying the variables that exist in the loss of inputs. With a fuzzy linear programming model, Vafaeenezhad and Tavakkoli-Moghaddam [19] improved planning and supply chain.

Due to the prevailing need for companies to efficiently utilize available resources, various studies with linear programming applications have been conducted in multiple cosmetic industries worldwide. The study by Aktar [2] for a cosmetics company considers the following restrictions for formulating an optimal cost of production: labor cost, cost of raw material, and machinery cost. The result of the study determined that the cost of raw materials is the most sensitive. Therefore, by reducing the cost of raw materials, the profit increases.

Given the importance of the supply chain in the cosmetics industry, the study by Velez-Gallego and Smith [20] used a Mixed Integer Linear Programming (MILP) model to solve the problem of product allocation in the quick-picking area of a cosmetics company. Through the mathematical model, labor efficiency was improved by concentrating picking activities in a more compact area, thus reducing the distance traveled by pickers.

A critical aspect that determines consumers' acceptance of cosmetic products is the product's quality. Generally, this aspect is determined by the components used and the effect at the sensory level, which the consumer perceives. Hathairat [6] used a linear programming model to optimize products taking as constraints the sensory attributes and the ideal product profile of a hand cream. Based on the results obtained, a hand cream was prepared using the optimal formulation, and the model was tested with a trained sensory panel. The results showed good acceptance by the consumer.

The literature review evidenced the application of linear programming for developing mathematical models, whose parameters were production costs, product sales prices, and inventory costs. Similarly, most mathematical models aim to maximize profits or minimize costs. Among the most common restrictions found in mathematical models are the restrictions of supply and demand, inventory, maximum capacity, and material requirements.

In Latin America, SMEs constitute the most economically productive sector of the countries. SMEs have a relevant role in the economic growth of Ecuador, and these represent about 94.7% of all companies in the country, generating about 70% of employment and contributing approximately 25% of the gross domestic product. Due to the importance of this economic sector for the country, these companies must manage high levels of productivity and quality [1].

The optimization of resources and budgets in companies such as SMEs is crucial for their success and competitiveness in the market. Poor production management can increase operating costs and the value of the final product, as well as limit the ability to offer competitive prices and discounts, which affects your presence in the market. Factors such as the cost of raw and packaging materials, determined by the supplier, impact operating costs. Choosing the right supplier is essential to improve production costs, and this choice must be based on a meticulous analysis of parameters such as the unit cost of materials, delivery time, and the minimum purchase batch. A robust vendor database facilitates this selection process, making it an analytical rather than an empirical process.

This scientific article aims to generate a linear programming model that reduces the operating cost of an SME-type company, focusing on optimizing raw material and packaging material costs, taking as decision variables the unit cost of raw material and packaging material, delivery time, and the minimum purchase lot. The mathematical model also considers the company's waiting time and storage policies. In this way, the model offers companies a practical option to optimize production costs by selecting the best supplier based on an analysis and comparison of the costs, times, and minimum purchase batches offered by each of them, allowing them to be more competitive in the market.

2 Methodology

2.1 Description of the Case Study

For the approach of the linear programming model that optimizes production costs in this scientific article, a product manufactured in an SME-type cosmeceutical company has been selected as a case study. The enterprise is an Ecuadorian company that produces personal care products like cosmetics, dermatological, and cosmeceuticals. Its portfolio includes antiseptic and moisturizing products. Some of its exclusive formulas contain N-Alkyl Dimethyl Benzyl Ammonium Chloride (BZK), a first-generation quaternary ammonium approved by the FDA, as a hand sanitizer to eliminate 99.9% of viruses and bacteria. The company has a production line for skin care, so all formulas are free of paraben preservatives. The company is geographically located in the north of Quito, Ecuador.

The company's current sales model is on demand and national scale; products reach customers through distributors. The company's production cycles are governed by customer demand for the product. The company also provides advisory services for formulating and obtaining sanitary registries and carries out maquilas of other personal care products.

The company is made up of two areas: administrative and operational. The administrative area is in charge of hiring personnel, purchasing raw materials, meeting with clients, monitoring sales, and developing new products. The operational area, for its part, is in charge of manufacturing the different products under quality standards. The company's operations are regulated through compliance with local regulations, it has operating permits, and its products have the corresponding health records. The company's production costs are mainly linked to the origin and minimum purchase quantity of raw materials and packaging, which can be from international or local suppliers. Occasionally, this creates an unnecessary inventory of these materials, which directly affects the warehouse space that the company has. Another factor that affects the cost is the minimum production batches related to the installed capacity of the equipment and the criteria for reducing direct variable costs. The latter is negotiated with the client in the case of maquilas. Figure 1 shows a scheme of the production and optimization process proposed in this chapter.

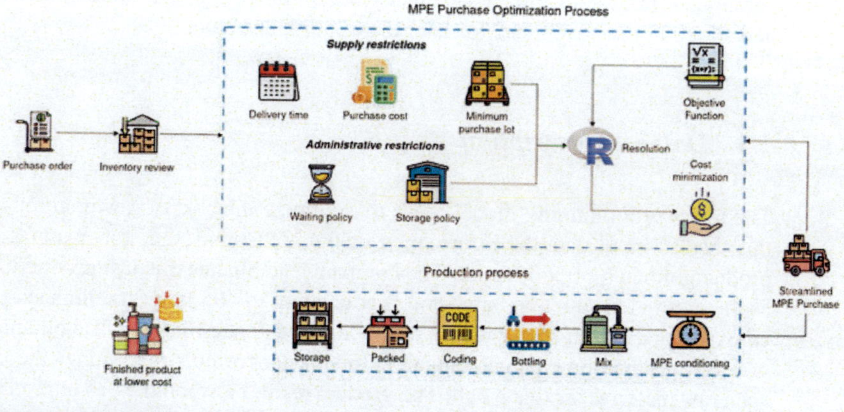

Fig. 1 Methodological scheme of the MPE purchase optimization process

2.2 Production Process

To establish the optimization model that minimizes the operating cost, it is necessary to describe the manufacturing process that the company manages for one of its products. Specifically, the case study of a 75g hand cream with antiseptic and moisturizing characteristics will be analyzed.

The manufacturing process begins with the customer's request for the product, which generates a purchase order; with this information, the available stock of the finished product in the storage warehouse is reviewed. The corresponding production batch is calculated if the need exceeds the available inventory. For this calculation, the amounts to be used for each Material and Packaging Material (MPE) are considered according to the master formula of the product. The costs relevant to this stage of the process are the Unit Purchase Cost (CUC) and the Minimum Purchase Lot (LMC) for each available Supplier (P).

Once the batch size has been determined, the production process begins with the review and approval of the MPE. Subsequently, the production order containing the exact quantities used in the production lot is issued. This order authorizes the enlistment of the MPE by weighing and counting each component. Mixing the elements in the reactor is necessary to transform the raw material into a finished product. During this stage, physical–chemical parameters are controlled to guarantee the quality of the product. When the product has reached the established quality parameters, it is transported to the semi-automatic packaging machine for the container.

Subsequently, when all the units of the manufacturing orders are obtained, the coding is carried out that will allow product traceability, and samples are sent to the laboratory to carry out the corresponding microbiological analyses and the subsequent release of the batch. Finally, the product is packed in cardboard boxes and stored in the warehouse for later distribution and sale.

2.3 Dataset Description

The linear programming model seeks to minimize the cost of raw material and, for this, has as input data the information of the MPE necessary for manufacturing a production batch of 5000 units. The company establishes this last production value as the optimal production batch and is equivalent to 405 kg of the finished product, considering a loss of 8%. The loss was calculated concerning the data obtained from previous productions. Its percentage in the master formula determines the needs of each raw material for this batch. The product result between the production batch in kilograms and the rate of each raw material is shown in Table 1.

The mathematical model also requires the suppliers' information, the company's waiting times, and storage policies. Ergo, the database of each raw material's

Table 1 Quantity of raw material in kilograms and units needed to manufacture a production batch of 5000 units

Component	Category	% Formula	MPE quantity	Drive unit (DU)	Max. quantity	(kg or und)
A	Solvent 1	7.2	29.2	Tank	2	540 kg
B	Moisturizing	6.2	25.1	Quintal	2	50 kg
C	Surfactant	5.7	23.1	Bucket	2	40 kg
D	PH regulator	6.2	25.1	Quintal	2	50 kg
E	Solvent 2	9.2	37.3	Tank	2	540 kg
F	Emulsifier 1	6.7	27.1	Quintal	2	50 kg
G	Emulsifier 2	6.0	24.3	Quintal	2	50 kg
H	Emulsifier 3	5.5	22.3	Quintal	2	50 kg
I	Preservative 1	4.5	18.2	Quintal	2	50 kg
J	Antibacterial	4.2	16.9	Bucket	2	40 kg
K	Preservative 2	4.5	18.2	Quintal	2	50 kg
L	Fragrance	4.8	19.2	Gallon	6	24 kg
M	Thickener	4.4	17.6	Gallon	6	24 kg
N	Solvent 3	25.0	101.3	Quintal	2	50 kg
O	Container	100	5000	Boxes	42	42 und
Q	Boxes	100%	50	Boxes	210	210 und

suppliers was collected since 2020; said information is hosted in the following repository: https://github.com/dievalhu/cosmetics_data.

2.4 *Optimization Model*

A Binary Integer Linear Programming (BILP) was used to select the best supplier of each raw material and packaging material, considering the following variables: waiting for time policies and company storage, defined as Supplier Purchase Cost (CC), Supplier Minimum Purchase Lot (LMC), Supplier Delivery Time (TE), Company Waiting Policy (PE), and Company Storage Policy (PA).

The methodological scheme shown in Fig. 1 summarizes the production process together with the variables and restrictions considered for the approach of the linear programming model. The terminology for mathematical modeling uses set theory. The set of components used in the manufacture of the product is represented by Eq. 1.

$$n \in n = \{A, B, \ldots, Q\} \tag{1}$$

The number of suppliers for each material is defined by Eq. 2.

$$i \in P = \{P_1, \ldots, P_i\} \tag{2}$$

Objective Function

The cost minimization function is mathematically expressed as

$$Z_{min} = MPE_A CC_A P_i + MPE_B CC_B P_i + \ldots + MPE_n CC_n P_i \tag{3}$$

This is,

$$Z_{min} = \sum_{n=A}^{n} MPE_n CC_n P_i \tag{4}$$

where:

Z = Objective Function
MPE_n = MPE Quantity
CC_i = Purchase Cost of the Selected Supplier

Definition of Constraints

Three constraints define the convex space on which an optimal solution of the linear programming mathematical model is sought. The restrictions are due to the organization's manufacturing process and administrative policies.

1. **Waiting Time Policy (PE)** The company has established that the ideal waiting time for the supply of MPE does not exceed three weeks. Since this time allows the company to be more flexible in its production process, this restriction is given by Eq. 5:

$$\sum_{n=A}^{n} TE_i P_i \leq PE \tag{5}$$

   where:

   TE_i = Delivery Time of each MPE
   P_i = Selected Provider (Decision Variable)
   PE = Company Waiting Policy = 3 Weeks

2. **Storage Policy (PA)** The finished product (PT), raw material, and packaging material (MPE) storage warehouse has limited capacity; therefore, it has been established, according to the layout and the current demand for the product, that the amount of raw material in the warehouse is the one summarized in Table 1.
3. **Minimum Purchase Lot (LMC)** The minimum purchase quantity is constrained by the amount needed to manufacture a production lot and the negotiation terms

with MPE suppliers. Usually, the unit cost of raw materials and packaging material is inversely proportional to the purchase volume, i.e., the higher the purchase volume, the lower the unit cost, applying economy of scale. However, the company cannot always acquire the largest batch at the lowest price since it has a storage policy according to its capacity. Equation 6 represents the restriction of the minimum purchase lot and considers the company's storage policy:

$$\sum_{n=A}^{n} LMC_i P_i \leq \sum_{n=A}^{n} PA_n \tag{6}$$

LMC_i = Minimum Purchase Lot from MPE supplier
PA_n = Storage Policy for each component

4. **Weighted Supplier Selection by Component** This restriction allows weighting the selection of each supplier by component, i.e., the sum of all suppliers always meets 100% of the demand the company requests. Thus, more than one vendor assigned to each component could result.

$$\sum_{j=1}^{i} P_i \leq 1 \tag{7}$$

However, since the historical data of the analyzed company have a single supplier per component, the model for selecting a single supplier P_i was forced. The objective of this last restriction is to compare, a posteriori, the pre- and post-optimization results. This implies that the model has a binary decision variable, and then the model is now Binary Integer Linear Programming (BILP).

Finally, the entire set of decision variables P_i are non-negative:

$$P_i \geq 0 \tag{8}$$

3 Results and Discussion

The proposed mathematical model was solved by using the *lpsolve* library in the R programming language. The results of the BILP model optimized the selection of a single supplier for each component and its corresponding unit cost, which made it possible to calculate the total cost of MPE according to the master formula of the product. It is important to note that this model can be modified to select multiple providers if the binary constraint on the decision variables is omitted. This feature is handy in larger companies with large purchasing volumes and multiple

Table 2 Comparative results of the selection by component in pre- and post-optimization

Component	Optimized CC	Optimized supplier	Unoptimized CC	Unoptimized supplier	Cost difference (CD)	% CD
A	$ 476	P18	$ 487	P7	$ 11	2%
B	$ 532	P5	$ 532	P5	$ -	0%
C	$ 241	P22	$ 262	P8	$ 21	8%
D	$ 101	P1	$ 104	P3	$ 2	2%
E	$ 280	P6	$ 280	P6	$ -	0%
F	$ 113	P7	$ 137	P8	$ 25	18%
G	$ 91	P5	$ 91	P5	$ -	0%
H	$ 104	P9	$ 108	P6	$ 4	3%
I	$ 53	P3	$ 55	P6	$ 2	3%
J	$ 112	P4	$ 213	P7	$ 101	47%
K	$ 116	P32	$ 123	P3	$ 7	5%
L	$ 850	P1	$ 1173	P2	$ 323	28%
M	$ 106	P8	$ 113	P1	$ 7	7%
O	$ 3800	P16	$ 4368	P11	$ 568	13%
Q	$ 59	P13	$ 63	P12	$ 4	6%

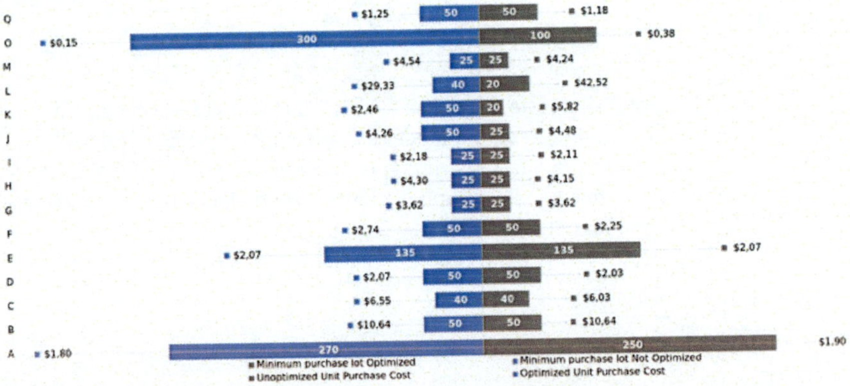

Fig. 2 Comparison of unit costs and LMC pre- and post-optimization by component

supplier options. Experimental calculations based on the information collected from the suppliers owned by the study company were used to validate the model.

The Supplier's Total Purchase Cost (CC) information by component is presented before and after optimization. The CC corresponds to the total acquisition cost of the LMCs, which satisfy the production requirement, multiplied by the CUC unit purchase cost of each selected supplier. Table 2 shows that the optimized selection of suppliers, through the mathematical model, reduces the total purchase cost of MPE by 13%, representing a saving of 1075 dollars for the company. Figure 2 shows the pre-optimization and post-optimization results for each component.

The most significant percentage savings, 47%, was achieved in component J. The manually selected provider (P7) offers double the LMC compared to the model-selected provider (P4). Therefore, the model prioritizes the supplier with the lowest total purchase cost due to the LMC constraint. The optimized selection of this component is an example of the comprehensive analysis of variables and constraints within the model compared to the empirical selection.

On the other hand, the lowest percentage of savings corresponds to components B, E, and G. In these cases, the supplier selected by the model is the same as the one chosen manually. In component B, the model prioritizes the company's waiting policy since only the selected supplier meets the delivery time within the established. For components E and G, the model prioritizes the choice of the supplier with a lower LMC and, therefore, lower purchase cost, even though the unit cost is higher than manual selection. Concerning components C, F, D, H, I, M, and Q, the model prioritizes providers offering lower unit costs where LMCs are equal to or less than those manually selected providers. Therefore, the optimized CC is lower than that of the manual selection.

Additionally, for component O, it can be seen that the manually selected supplier offers a lower unit purchase cost. However, it also provides a 200% higher LMC compared to the provider chosen by the model. In this case, acquiring the materials from the supplier with a high LMC generates a high inventory cost. It considerably reduces the storage capacity of other raw materials. These are the main reasons the company decided to implement a storage policy. The heading of this component represents a considerable investment because the LMCs offered by all providers are more significant than the rest of the components. In this sense, it may be helpful to add an investment limit restriction to the model to establish the maximum amount to invest in each component or the total purchase. This can be especially useful for SMEs with limited investment capital.

In addition to the decision variables taken into account in this model, others can be included to improve the results, including additional characteristics of the provider to have a more holistic vision in decision-making. A model that includes other decision variables is proposed by the authors Torğul and Paksoy [18], who propose a multi-objective fuzzy linear programming model for selecting ecological and efficient suppliers. For this, the authors included two objective functions: the first that maximizes the ecological value *green* and the second that maximizes the providers' efficiency value *lean*. The authors used demand constraints, carbon emission capacity limit, lead time, and budget. Then, the model proposed in our research work can reference the study carried out by Torğul et al. to weight variables, add restriction criteria, and test providers with a prior weight. Given that the proposed BILP model is scalable, the integrated values of the weighting for each supplier could be added to the current objective function, keeping the same restrictions. The result would minimize the purchase cost of MPE considering the ecological and efficient factors.

Although this scientific article considers three decision variables, unit purchase cost, minimum purchase lot, and delivery time, the scalability of the model allows for the incorporation of new variables a posteriori. The review of the scientific literature indicates that the model could consider additional variables such as maximum investment capital, payment agreements, cumulative and incremental discounts, product quality, storage costs, stochastic fluctuations in purchase prices, service, and customer experience, purchase, and the socioeconomic and environmental impact of the product offered.

4 Conclusions and Limitations

The results obtained in this chapter allowed us to generate a 13% reduction in the company's total purchase cost, representing 1075 dollars. These results show that it is possible to optimize the operating cost of a company through BILP by proposing as an objective function the minimization of the total purchase cost of raw materials and packaging material, considering the decision variables unit purchase cost, minimum purchase lot, delivery time, and company policy restrictions. The proposed BILP model is flexible and adapts to the needs of different types of companies in selecting suppliers. This model can be modified to allow the selection of several suppliers simultaneously, which is especially useful in situations of shortage of MPE and in large companies with a more extensive supplier database. In addition, the flexibility of the model allows for adding criteria in the selection of suppliers, which is essential for managing the supply chain in companies. In particular, the BILP model is a suitable solution for supplier selection in SMEs with limited resources, where supplier selection is crucial.

References

1. V. Aguiar, A. Arghoty, S. Burgos, M. Gualavisí, M. Onofa, P. Ruiz, M. Sáenz, R.G. Jaramillo, L.M. Carrasco, V. Sión, et al., *Estudios industriales de la micro, pequeña y mediana empresa* (FLACSO Sede Ecuador, Ecuador, 2013)
2. S. Aktar, Linear programming problem formulation and solution using benders' decomposition method (2020)
3. A. Alvarado, Tendencias cosméticas cambian por la pandemia (2020)
4. D.A. Collier, J.R. Evans, *Operations and Supply Chain Management* (Cengage Learning, 2020)
5. G. Dantzig, Linear programming and extensions, in *Linear Programming and Extensions* (Princeton University Press, Princeton, 2016)
6. U. Hathairat, Linear programming and consumers' ideal sensory attributes in product optimization (1990)
7. J. Heizer, B. Render, *Principles of Operations Management* (Pearson Education, 2014)
8. G. Jones, Globalization and beauty: a historical and firm perspective. EurAmerica **41**(4), 885–916 (2011)
9. D. Kligman, Cosmeceuticals. Dermatol. Clin. **18**(4), 609–615 (2000)
10. S. Kumar, Exploratory analysis of global cosmetic industry: major players, technology and market trends. Technovation **25**, 1263–1272 (2005). https://doi.org/10.1016/J.TECHNOVATION.2004.07.003

11. A.L. Mafra, F.N. Castro, F.d.A. Lopes, Investment in beauty, exercise, and self-esteem: are they related to self-perception as a romantic partner? Evol. Psychol. Sci. **2**(1), 24–31 (2016)
12. A. Murillo, Estudio del efecto socio-financiero de la Industria Cosmética en la ciudad de Quito durante el año 2019. B.S. thesis, PUCE-Quito (2021)
13. C. Nitesh, V. Himanshu, D. Roshan, Cosmetics market by category (skin and sun care products, hair care products, deodorants & fragrances, and makeup & color cosmetics), gender (men, women, and unisex), and distribution channel (hypermarkets/supermarkets, specialty stores, pharmacies, online sales channels, and others): global opportunity analysis and industry forecast, 2021–2027 (2021)
14. V. Ortiz, Álvaro Caicedo, Plan óptimo de producción en una planta embotelladora de gaseosas. Revista Ingeniería Industrial **11**, 69–82 (2012)
15. K. Randheer, Optimization-maximization of marketing information systems in small and medium enterprises: a linear programming approach. Int. J. Bus. Soc. Sci. **5**(7), 102–112 (2014)
16. R.R. Tan, K.B. Aviso, C.D. Cayamanda, A.S. Chiu, M.A. Promentilla, A.T. Ubando, K.D. Yu, A fuzzy linear programming enterprise input–output model for optimal crisis operations in industrial complexes. Int. J. Prod. Econ. **181**, 410–418 (2016). https://doi.org/10.1016/J.IJPE.2015.10.012
17. A. Tedeschi, L.E. West, L. Guzzardi, K.H. Bhatt, E.E. Reid, G. Scapagnini, G. Micali, Cosmeceuticals in dermatology, in *Update in Cosmetic Dermatology* (2013), pp. 87–113
18. B. Torğul, T. Paksoy, A new multi objective linear programming model for lean and green supplier selection with fuzzy TOPSIS, in *Lean and Green Supply Chain Management: Optimization Models and Algorithms* (2019), pp. 101–141
19. T. Vafaeenezhad, R. Tavakkoli-Moghaddam, Fuzzy multi-objective supply chain master planning in a wood and paper industry: a case study. IFAC-PapersOnLine **49**, 1632–1637 (2016). https://doi.org/10.1016/J.IFACOL.2016.07.814
20. M. Velez-Gallego, A. Smith, Optimization of a fast-pick area in a cosmetics distribution center, in *Progress in Material Handling Research 15th* (2018)
21. A. Villa, T. Taurino, Open-loop-feedback scheduling for multi-stage production systems. Proc. CIRP **33**, 99–104 (2015). https://doi.org/10.1016/J.PROCIR.2015.06.019

Part II
Sustainable Communications & Information and Communications Technology

Data Collection with Honeypot Server for Reverse Engineering of Malware

Ivan Cvitić, Marko Periša, and Josip Vladava

1 Introduction

There is an increase in the use of technology, primarily the Internet, which inherently leads to cybercrime that can threaten anyone. Victims can be individuals, organizations, companies, government bodies, and governments themselves. Cybercriminals use malware to steal sensitive data or financial resources, cause damage, take remote control of other computers, or spy. Malware is any harmful code that affects hardware, software, a network, or the user.

In response to the increased development of malware, the field of malware analysis is emerging. Malware analysis is the process of discovering the source, functionality, and impact of malware using reverse engineering methods and techniques. The analysis is carried out after a security breach in order to gather information regarding how the malicious program entered the system, what damage it causes and how to remove the malicious program from the system, and ultimately, how to secure the system from similar threats in the future. In order for reverse engineering tools to keep up with the evolution of malware, there is a need for honeypot servers through which new malware samples are collected. Honeypot servers serve as a trap to catch cybercriminals. Through such servers, researchers gain insight into the activities that attackers – cybercriminals – carry out after gaining access to someone else's computer.

Research in this paper is based on data collected by connecting to an implemented honeypot server and analysis of collected data. Research is focused on the analysis and reverse engineering of gathered malware samples. Based on the

I. Cvitić (✉) · M. Periša · J. Vladava
University of Zagreb, Faculty of Traffic and Transport Sciences, Zagreb, Croatia
e-mail: ivan.cvitic@fpz.unizg.hr; marko.perisa@fpz.unizg.hr; josip.vladava@fpz.unizg.hr

L. Knapčíková, D. Peraković (eds.), *8th EAI International Conference on Management of Manufacturing Systems*, EAI/Springer Innovations in Communication and Computing, https://doi.org/10.1007/978-3-031-53161-3_5

results of the analysis, guidelines for improving the security of information and communication systems are presented.

This paper is organized as follows. The second section contains an overview of previous research related to the field of reverse engineering of malware and an overview of honeypot servers. The third section describes the characteristics of honeypot servers, which are divided according to the level of interactivity and purpose. This section also contains an overview of the most commonly used methods, static and dynamic analysis, which are used in the process of reverse engineering of malware, through which an insight into the way a malicious program works is provided. The fourth section serves to analyze the data collected through the implemented honeypot server. The collected data is divided into connection data, downloaded scripts, and downloaded malware. The fifth section provides guidelines for improving server security based on the synthesis of the results obtained from the previous analysis.

2 Previous Research

The field of malware development is fast-growing, malware developers are using intricate tools and techniques to find exploitable vulnerabilities, obfuscate malware, and evade antimalware software. To counteract increasingly complex malware, it is necessary to stay up-to-date with current malware attacks. By combining honeypot servers and tools and techniques of malware reverse engineering, it is possible to gather and analyze current malware attacks delivered through poorly configured web servers. By analyzing these attacks, it is possible to gain insight that can be used to increase the security of web servers and their services. Statistical data from 2022 show that in 2020, there were 5.6 billion attacks by malware worldwide. Most of those attacks were aimed at the professional sector of North America. A large number of attacks in 2020 are attributed to the rise in the prices of cryptocurrencies, which led to the increased development of cryptocurrency stealing malware or malware that is designed to use the resources of infected computers to carry out cryptocurrency mining activities. Most organizations were attacked by botnets and cryptocurrency mining malware, while ransomware attacks accounted for only 5% of total attacks worldwide. Furthermore, 85% of organizations were found to be able to recover data without paying a ransom. In 2021, 5.4 billion attacks with malicious programs were detected, and in the first half of 2022, 2.8 billion attacks were detected, which is a significantly lower number of detections compared to 2018, when 10.5 billion malware attacks were detected on a global level [1–3].

Previous research can be divided into honeypot servers research and research on methods and techniques of reverse engineering malicious programs. The research presented in [4] provides an insight into honeypot servers, their importance for network security, advantages, disadvantages, their level of interactivity, and the risks associated with them. Honeypot servers would not have any significance if the attacker did not interact with them; therefore, they are divided according to the level of interactivity, which can be low, medium, or high. Honeypot servers of low interactivity only provide services such as FTP (File Transfer Protocol), HTTP (Hypertext Transfer Protocol), and SSH (Secure Shell); they have the role of passive monitoring of connections, and are accompanied by low risk. An example of low interactivity honeypot is Honeyd. Honeyd is an open-source honeypot, and it provides the ability to create several virtual devices through unused network IP addresses. Medium interactivity honeypots are similar to low-interactivity honeypots but have a medium level of risk, collect more detailed information about requests and they can gather malware samples. Examples are Dionea and Napenthes. Unlike medium and low-level ones, high-interactivity honeypots include the actual operating system, are accompanied by high risk, collect all kinds of information, and are difficult to maintain; an example of such a honeypot is Spectre. In addition to the level of interaction, we also divide them according to their purpose into research and production honeypots. According to the authors of this paper, some of the advantages of honeypot servers are the ability to catch malicious activity, even if it is in encrypted form, and improve the intrusion detection system by reducing the number of false positives; honeypots work in any IP (Internet Protocol) environment, including IPv6 (Internet Protocol version 6), and require minimal resources. According to the authors, the disadvantages of honeypots are the risk created by attracting attacks if the honeypot is placed in the organization's environment and the lack of information if the attacks do not materialize [4].

The research presented in [5] uses a sample of the TT.exe malicious program and describes two main methods of analyzing malicious programs: static and dynamic analysis, each of which is divided into basic and advanced analysis. The main difference between the two is that static analysis is performed without executing the malware, while dynamic analysis is performed in an isolated secure environment where every change that malware makes is monitored. Basic static analysis consists of scanning the malware with an antivirus program, generating a unique hash value, detecting packers, and analyzing the PE (Portable Executable) format of the malware. Examples of tools that perform basic static analysis are VirusTotal, Md5deep, PEiD, Exeinfo PE, RDG Packer, and D4do. Advanced static analysis involves the analysis of strings from which the functions used in the code of the malicious file can be read, and the analysis of the file in the disassembly tool, which analyzes the assembly instructions from which the functionality of the program can be read. Examples of advanced static analysis tools are BinText, Dependency Walker, and IDA (Interactive Disassembler). Basic dynamic analysis consists of monitoring of running processes and network traffic analysis. Examples of basic dynamic analysis tools are VirtualBox, Anubis, Comodo, Instant Malware Analysis, and Wireshark. Advanced dynamic analysis is carried out using the debugger. Information about the characteristics of the specific malware is obtained from the results of the analysis. Examples of advanced dynamic analysis tools are OllyDbg and Regshot. This paper presents the analysis of the malicious program and concludes that it is necessary to use a combination of both static and dynamic analysis tools to present a complete picture of the malicious program [5].

Paper [5] studies static and dynamic analysis but focuses on the methods and tools of these analyses. Some of the mentioned tools are PEiD, PE view, PEExplorer, UPX (Ultimate Packer for Executables), Dependency Walker, OlldyDbg, Process Monitor, and Process Explorer. This paper also contains a comparison of Internet-based solutions for analyzing malware samples such as Any Run, Virus Total, Intezer Analyze, and HybridAnalysis. The conclusion of this paper gives insight into a simplified approach to the analysis of malware. An analysis diagram was created according to which static analysis is first conducted, followed by reverse engineering phase, and dynamic analysis phase. Each phase uses appropriate tools and techniques. Based on the gathered information or lack thereof, machine learning analysis can be conducted. The conclusion of this paper highlights the need to use more tools when analyzing a particular malicious program [6].

This research aims to fill out the research gap identified in previous research. In doing so, the data collected through the honeypot server was analyzed. The conducted analysis is divided into the analysis of connection data, entered commands, downloaded scripts, and downloaded malware. Based on the analyzed results of the conducted research and the conclusions reached, guidelines for improving the security of servers based on web server services are proposed.

3 Honeypot Server Characteristics in the Role of Collecting Malware for Reverse Engineering

One of the earliest mentions of the term honeypot dates back to the year 2003, honeypot was defined as: *a decoy computer resource whose value lies in being probed, attacked or compromised* [7]. Honeypots are categorized according to their purpose and level of interactivity. According to the purpose, they are divided into production and research honeypots, and according to the level of interactivity there are honeypots of low, medium, and high interactivity [4, 8]. Medium and high interactivity research honeypots are used to collect malware samples since they have the capabilities to safely download and store malware.

Production honeypots are used by large companies and organizations. Production honeypots are most often used to distract attackers from critical, higher-value infrastructure. Information such as the attacker's IP address, attack time, traffic volume, and targeted ports are collected through production honeypots. In this case, said information is not used for detailed analysis, but for efficient blocking of collected IP addresses and fixing security flaws. Production honeypots collect less data than research ones, and are easier to implement, maintain, and represent a lower security risk for the organization [4].

Research honeypots are complex, more difficult to implement and maintain, and can pose an increased security risk. They are most often used by military organizations, government bodies, and research organizations. Such honeypots are used to collect information about activities, intentions, and ways of achieving unauthorized access to the system, that is, they are designed to collect information about specific methods used by attackers, their behavior within the system, and detection of active exploits [4]. The above information is used to develop tools, programs, and processes for detection, mitigation, and analysis of attacks.

High-interaction honeypots are designed to make attackers invest as much time as possible inside the honeypot, thus the researchers or security team gather a large amount of data about the attacks, intentions, and attempts to exploit the system. High interactivity honeypots provide attackers with a real system to conduct the attack, thus reducing the likelihood that attackers will be able to detect the honeypot and change their behavior [8]. The information collected through this kind of honeypot provides a more detailed insight into the attacker's tools for gaining privileges in the system and the activities of gathering information and their exfiltration. In order to reduce the possibility of spreading attacks from the honeypot itself, it is necessary to provide a separate environment in which the honeypot is located [4].

Medium-interaction honeypots try to combine the best elements of low and high-interaction honeypots. They enable the collection of data of the activities the attackers carry out in the system, and the malicious programs that they leave on the system. Medium-interaction honeypots do not provide attackers with access to the real operating system, and they enable fake services through which attackers establish a connection [4]. Such honeypots are successful because they can be interesting enough to the attackers and provide administrators with a high level of configurability. Such honeypots are often used to collect data on the behavior of automated attacks and botnets.

Low-interaction honeypots usually provide the capability to imitate services such as SSH, FTP, or HTTP servers; they do not contain any files that could be of interest to attackers. Such honeypots are easy to implement and serve to collect information about IP addresses, attacker locations, port scanning attempts, and depending on the mimicked service, used usernames and passwords [4].

Medium-interaction honeypots provide the best characteristics for reverse engineering of malware. Reverse engineering of malware provides insight into how the malicious program functions, and how to stop it from operating and further spreading. Analysis of malicious programs is carried out using static and dynamic analysis methods and their associated tools. The static method represents the analysis process in which the malicious program is not executed, while the dynamic analysis implies the execution of the malicious program within a controlled environment and the collection of information regarding the activities of the malicious program.

3.1 *Static Analysis of Malware*

As already mentioned, static analysis is the first step when analyzing a malicious program. During static analysis, the malware is not executed, but an attempt is made to gain insight into basic information about the program, such as character strings that indicate malicious behavior, file format, used libraries, memory layout of program, hash value, and used obfuscation methods [9, 10]. The goal of static analysis is to find [the main characteristics of a particular malware so that it can be identified in the future. Techniques that can be used during static analysis are finding character strings, finding a unique key, scanning with antivirus programs, unpacking, PE file header analysis, ELF (Executable and Linkable Format) file header analysis, and disassembly [9, 11].

The most common first step in static analysis is string discovery, which involves detecting meaningful character strings within a malicious program's executable file. Via character strings, information about used functions, contacted IP addresses and domains can be obtained. Opening the executable file with a text editor will result in an unreadable string of characters but using the "strings" command on a Linux operating system or the Strings tool on a Windows operating system will provide insight into all character strings. Attackers can add character strings that the program ignores to make analysis difficult [11]. The second most common step is disassembly, which is applied to an executable file that contains a binary representation of the machine code. The disassembly process converts that binary code into a form that is understandable to humans, that is, into assembly language [12]. Commonly used disassembly tools are IDA Pro, Ghidra, Binary Ninja, and Hopper. Before moving on to dynamic analysis, it is necessary to use the other previously listed static analysis techniques in order to get a more complete picture of the analyzed malicious program.

3.2 *Dynamic Analysis of Malware*

Dynamic analysis is performed when all possibilities of static analysis have been exhausted and when all information that can be obtained through static analysis has been collected. Dynamic analysis observes the behavior of the malicious program during and after its execution. Using the dynamic analysis, an insight into the actual functionalities of the malicious program is gained, given that the functionalities detected by static analysis do not necessarily have to be performed. The activities performed by the malicious program may also depend on the environment in which it is executed. This kind of analysis is performed in a closed system, usually a virtual machine, in order to stop the spread to other computers or the network [6, 9]. During dynamic analysis, caution is required because there are malicious programs that can detect virtual machines and hide essential functionalities. There are also malicious programs that exploit vulnerabilities in virtualization software and manage to infect the host computer, but such attacks are rare. Techniques used when performing dynamic analysis are [9]: process monitoring, registry monitoring, file system monitoring, network traffic analysis, sandbox analysis, and debugger analysis.

The technique that best encompasses most of the dynamic analysis techniques listed is analysis in an isolated environment – sandbox analysis, which is used to prevent the uncontrolled spread of malware. Isolated environments for the analysis of malicious programs have similarities with the process of virtualization, but these are focused on the analysis of programs; therefore, they have additional functionalities such as creating a basic report, taking snapshots during the execution of malware, collecting network traffic, monitoring processes, monitoring the file system, and analyzing malicious IP addresses that malware contacts [9, 11]. A complete dynamic analysis can be carried out through solutions that offer isolated environments. Such solutions can be installed directly on the computer intended for analysis or an Internet-based service can be used. The solutions that need to be installed are Remnux, Windows Sandbox, and CuckooSandbox. Solutions available online are Any Run, Joesandbox, Hybrid Analysis, and Hatching Triage.

4 Analysis of Data Collected Through Honeypot Server

For the purpose of this research, a hybrid experimental platform was established in the Cloud environment in the Laboratory for Security and Forensic Analysis of Information and Communication System at the Faculty of Transport and Traffic Sciences, University of Zagreb. Data were collected via Cowrie medium-interaction honeypot server that was active for 96 hours. Honeypot was deployed via the DigitalOcean cloud service provider. Collected data can be classified into the following categories: connection data, executed commands within the system, downloaded program scripts, and downloaded executable files. Inside the system, attackers have the ability to download files. Downloaded files are mostly bash program scripts, ASCII (American Standard Code for Information Interchange) text documents, compressed files, or executable files. By downloading files, attackers test the availability of downloads within the server, which helps in the detection of honeypots, so the content of downloaded files can be test data in text form, HTML (HyperText Markup Language) code, or files used for exploiting the system, for example, malicious programs and bash scripts to download other malicious programs.

4.1 Analysis of Connection Data

In a 96-hour period, there were 7155 successful and 6415 unsuccessful connection requests made on the honeypot. The total number of generated events is 165,326. Location analysis of the connection data shows that the honeypot server had connection requests with IP addresses from 67 different countries around the world. The largest number of unique IP addresses came from the USA, 201 unique IP addresses from the USA tried to connect to the honeypot. All other countries have 55 or fewer unique IP addresses. Figure 1 shows the top ten countries by the number of unique IP addresses that interacted with the honeypot. The total number of unique IP addresses is 822, which means that addresses from the USA represent 24.45% of the collected IP addresses.

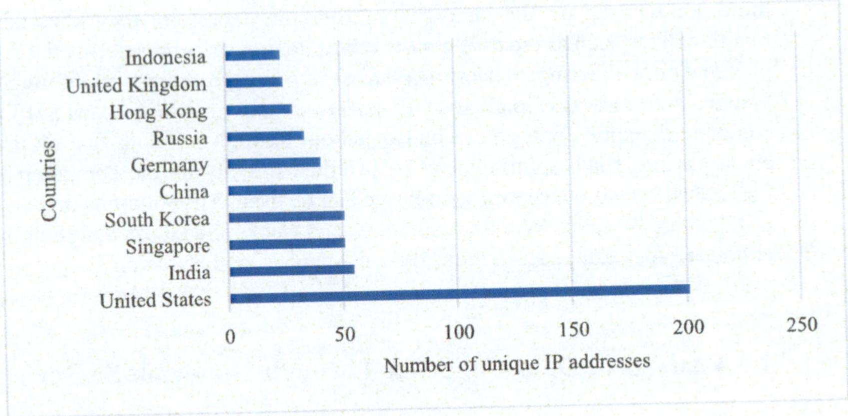

Fig. 1 Top ten countries by number of unique IP addresses

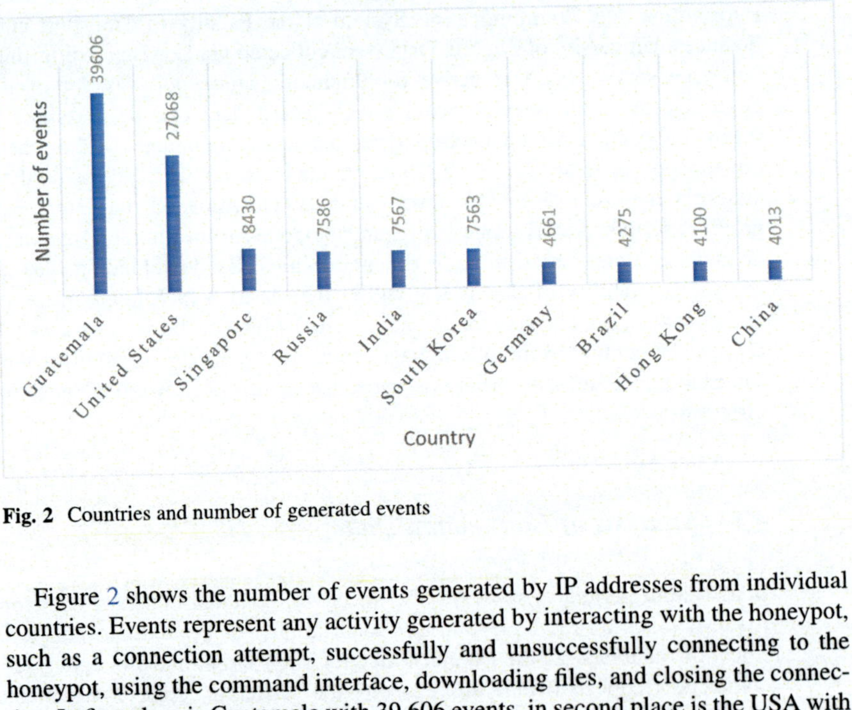

Fig. 2 Countries and number of generated events

Figure 2 shows the number of events generated by IP addresses from individual countries. Events represent any activity generated by interacting with the honeypot, such as a connection attempt, successfully and unsuccessfully connecting to the honeypot, using the command interface, downloading files, and closing the connection. In first place is Guatemala with 39,606 events, in second place is the USA with 27,068 events, while all other states have less than 8500 generated events.

Detailed analysis of the IP address from Guatemala shows the pattern that generated such a high number of events. The following events were generated by said IP when trying to connect to the honeypot: detection of client SSH service version, connection of client and honeypot, successful connection using root as

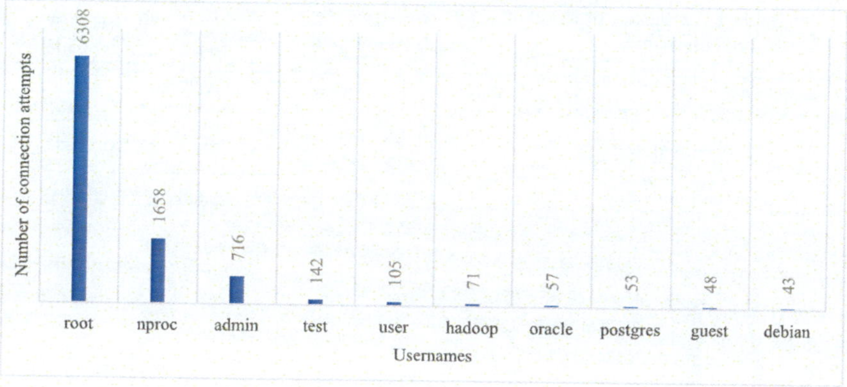

Fig. 3 Ten most used usernames

username and random password, entered command "echo -e \x6F\x6B," command response generated, and closing of the connection by the client. In this way, 39,606 events were generated within one hour. Most of the other events follow a similar pattern, that is, a large number of generated events in a short period of time, which points to the fact that there are a large number of automated programs that scan the Internet trying to find poorly configured devices. Similar to the example of the Guatemalan IP address, the events generated by the US IP address point to a more sophisticated automated program that, after a successful connection to the honeypot, enters commands to try to detect whether it is connected to a virtual machine.

Honeypot server gathered information about interactions with the ports. Information about ports consists of the sum of successful and unsuccessful connections to ports, and communication achieved by scanning logical ports. Using this information, it is possible to see that the most frequently used port is the SSH port. Other ports had a significantly lower number of interactions.

The total number of gathered usernames is 2121, the most common usernames collected use the names of frequently used services and protocols, such as FTP, Oracle, Debian. The least frequently used usernames consist of random numbers or represent human names, for example, 11111, Amy, 123qwe, Akash, Alberto. Top ten most frequently used usernames are shown in Fig. 3.

The honeypot server is configured to allow successful connection only with usernames from Table 1 with any combination of passwords. Table 1 shows that most successful connections were made using root and admin usernames.

Along with usernames, 8770 different passwords were collected. Figure 4 contains ten most used passwords. First place is the password nproc, which was used 1658 times. Nproc is a command within the Linux command interface that is used to print the number of processor units available in the system or the current process. Although it is in the first place, this password cannot be found in the most used wordlists, while other collected passwords can be found.

Table 1 Frequency of use of allowed usernames

Username	Number of uses
root	6308
admin	716
user	105
server	20
administrator	16
database	2

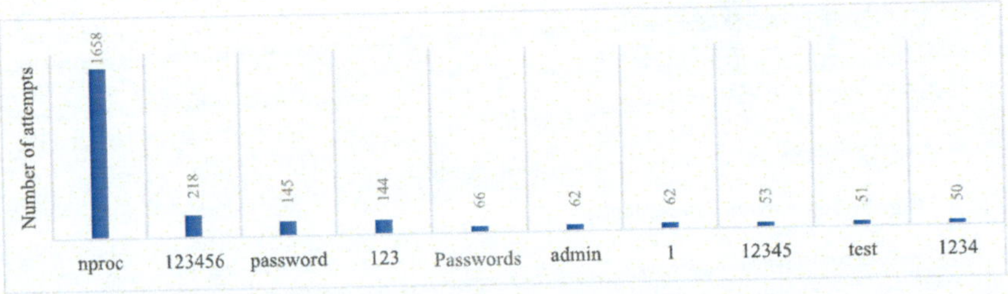

Fig. 4 Ten most used passwords

Table 2 Ten most frequently used commands

Command	Number of entries
echo -e "\x6F\x6B"	4949
uname -a	1766
free -m | grep Mem | awk '{print $2,$3, $4, $5, $6, $7}'	1734
cat /proc/cpuinfo | grep name | head -n 1 | awk '{print $4,$5,$6,$7,$8,$9;}'	1734
which ls	1733
w	1733
uname -m	1733
uname	1733
top	1733
ls -lh $(which ls)	1733

Collected commands are divided into most and least used. Table 2 shows the most frequently used commands, the command used by the IP from Guatemala occupies first place, the analysis of that command leads to the conclusion that it is not very useful. The command is used to print the character string "ok." Other commands in Table 2 are used to retrieve information about the processor, file system version, machine name and version, and processor architecture. Using these commands, the attacker can find out if it connected to the virtual machine.

Collected rarely used commands are also used to gather information about the processor. But other commands, such as "wget" command is used to download a file from a remote server, in the continuation of the command, the option to start the downloaded script is added, which is then started. Other rarely used commands are mostly used for changing the password within the command line interface.

Table 3 Characteristics of collected malware samples

	Sample1	Sample2
MD5	63d6cd74a7cd01bf3a3921c36e90237f	262319f550cc09ccd489f1caf254e54b
SHA1	f697783da228c7787cf1c6a67a10a8 c065d6aaa7	243b1043c72ce76aaefa1c84b39b00 778ae1b53f
SHA256	4f02cc4d5426b63e3eca3ada3c9a8a 111a952c0e373c5500519ea8eea5ade853	ad2d2ae296c85792794bdf2d77efa5 f56d07846f091037661392c697febaebb8
File size	549 KB	10.9 MB
File type	ELF 32-bit	ELF 64-bit

4.2 *Analysis of Scripts*

Bash scripts make up 362 of 366 downloaded files. These scripts share nearly identical code, which is used to download other malicious files. Execution of these scripts can be described as follows:

1. Determining the file path where the file will be downloaded, such as /tmp, /var/run, /root, or /home.
2. TFTP (Trivial File Transfer Protocol) protocol is used to remotely download malicious files.
3. Content of the malicious file is copied into a new file called "badbox."
4. Newly created "badbox" file is given execution privileges.
5. "Badbox" is executed.
6. Steps are repeated, but the malicious file from step 2 is changed so it targets different CPU architecture.

As it was mentioned in step 6, bash script is used to download malicious files that are targeting different CPU architectures, such as MIPS, MIPSEL, SH4, x86, ARMv6, i868, PowerPC, i586, m68k, SPARC, ARMv4, and ARMv5. These bash scripts have nearly identical code which only differs in IP addresses that are contacted to download the malicious files. After analyzing bash scripts, only three unique IP addresses were identified. These IP addresses represent the server that hosts malicious files.

4.3 *Malware Analysis*

The analysis of the collected malware samples will be carried out using reverse engineering tools and techniques. Malware samples were collected through Cowrie medium-interaction honeypot, their names, sizes, file types, and hash values are shown in Table 3.

During the analysis of Sample1, it was revealed to be a botnet malware named XORDDoS. The name of this malware was obtained by detection through

VirusTotal site, manual static and dynamic analysis of this sample did not yield any significant information; therefore, the process of reverse engineering of the Sample2 is described in more detail.

More information was obtained by analyzing Sample2. Initial static analysis of the Sample 2 ELF header determined that it is a 64-bit ELF file intended for the x86-64 processor architecture. Through strings analysis it was determined that Sample2 is a script written in the Python programming language, which was converted into an ELF executable using PyInstaller. The Detect It Easy tool confirms that it is indeed a packed program, but the program used for packaging was not successfully detected. The tool did not recognize the packer because PyInstaller was used, which works in a similar way to packer programs. In simplified terms, the original python script is encapsulated with the ELF format, and in this way, the script becomes an executable file. In order to get more information about Sample2, it is necessary to transfer the pydata section of the file to another file called pydata.dump, this new file was extracted using the Pyinstxtractor tool. After unpacking, files with the extension ".pyc" were obtained, on which the decompilation process is carried out using the Uncompyle6 tool. After these steps, a Python script containing the source code of the malicious program was successfully obtained. Before analyzing the program code, a dynamic analysis was performed on Hatching Triage. Analyzing the network activity, using the Wireshark tool, it is possible to see that Sample2 is making DNS requests looking for a domain on the Tor network. The infected computer then successfully opens a TCP connection with the obtained IP address of the server within the Tor network, but due to a faulty TLS (Transport Layer Security) certificate, the connection is closed. Through traffic analysis, malicious IP addresses and domains were found. Analyzing the processes through the Joesandbox resulted in a list and hierarchy of processes initiated by Sample2. One of the processes creates a file in the "/tmp/" directory, the content of the file is Sample2 itself, and the location "/tmp/" is chosen because the malicious program does not need to have permission to access that directory. Another interesting process executes the "uname -p" command through which malware checks the CPU architecture. With additional process analysis, one can notice a larger number of processes through which information about the CPU is found. Other processes are used to activate the "cron" service and shut down the processes that the malicious program initiated.

Analysis of the malware's source code revealed that the Python script contained code that could be run on Windows and Linux operating systems. Sample2 first identifies the architecture of the system it is running on, CPU information, and other system details. During execution, sample2 checks if there is a "kinsing" process running on the system; if there is, it shuts down. Kinsing is a malicious botnet program. After that, it locates and deletes the existing "cron" configuration and sets up a new "cron" task that ensures that the malware will be launched every time the system is booted. Mentioned functionalities are shown through the code in Fig. 5.

The main function of Sample2 is cryptocurrency mining, specifically, Monero cryptocurrency. Mining of the Monero cryptocurrency is performed using a processor. In the case of executing Sample2 on the Linux operating system, the program may use a maximum of 75% of the processor's resources, while this limit is set to

```
def crontab():
    try:
        subprocess.Popen('crontab -r', stdout=(subprocess.DEVNULL), stderr=(subprocess.DEVNULL), shell=True)
    except:
        pass

def kinsing():
    try:
        subprocess.Popen("kill -9 $(pgrep -f 'kinsing');kill -9 $(pgrep -f 'kdevtmpfsi');rm /tmp/kinsing;rm /tmp/kdevtmpfsi;",
        stdout=(subprocess.DEVNULL), stderr=(subprocess.DEVNULL), shell=True)
        subprocess.Popen('echo no > /tmp/kinsing; chmod --------- /tmp/kinsing; echo no > /tmp/kdevtmpfsi; chmod --------- /tmp/kdevtmpfsi;',
         stdout=(subprocess.DEVNULL), stderr=(subprocess.DEVNULL), shell=True)
    except:
        pass
```

Fig. 5 Code snippet of Sample2 for maintaining persistence

25% on the Windows operating system. Mined cryptocurrencies are then sent to a specified wallet.

The "startHider" function is used to hide the malware on Windows systems. If the Task Manager is opened the mining activity will be halted and resumed after closing the Task Manager. Other functionalities of this malicious program show Sample2 can also be used to carry out DDoS attacks.

5 Synthesis of Research Results and Protection Guidelines

5.1 Result Synthesis

Honeypot server was implemented as a part of the research. Honeypot was used to collect and analyze various data. Table 4 shows the types of performed analyses and their results. Guidelines for improving the security of information and communication systems can be presented based on the conducted analyses.

The analysis of the connection data showed that the largest number of unique IP addresses came from the USA, and the largest number of generated events in interaction with the honeypot was the IP address from Guatemala, which is the only collected IP address from that country. The data shows that SSH port had the most interactions, attackers target this port because they can get access to the command interface through which they can perform further system exploitation activities. Other ports that the attackers interacted with represent frequently used services such as: HTTP, HTTPS, and SMTP.

Analysis of collected usernames and passwords showed that attackers use publicly available wordlists of the most frequently used usernames and passwords. During the analysis, not a single example of a brute-force attack was discovered to find the correct combination of credentials, which means that attackers are increasingly relying on specified lists of credentials to avoid time-consuming attacks.

Used commands were collected after an attacker was successfully connected to the honeypot. Most commands are used to gather information about the processor and operating system architecture. Based on this information, it can be confirmed

Table 4 Result synthesis of collected data

Type of analysis	Collected data	Results
Analysis of connection data	IP addresses Targeted ports Generated events Usernames Passwords	Sources of attack come from all over the world SSH port has the most interactions Attacks were carried out using wordlists. Brute-force attacks were not detected Attacks were automated
Analysis of entered commands	Commonly used commands Rarely used commands	Most of the commands are used to gather information about the processor and operating system architecture Automated attacks can detect a virtualized environment
Analysis of scripts	*Bash* scripts	Script is used to download malicious programs designed for different CPU architectures Malicious IP addresses were detected
Malware analysis	Sample1 Sample2	Collected malware samples were reverse engineered using appropriate tools Identified techniques used for maintaining persistence Identified malicious IP addresses and domains

that the attackers create automated scripts that have the ability to detect virtual machines. If the virtual environment is detected, the scripts can terminate further execution or simulate activities that are performed to make analysis more difficult.

The attackers downloaded 362 bash scripts to the honeypot, their analysis showed that they were used to download other malicious programs to honeypot. The scripts allow downloading programs for different CPU architectures and executing them. By analyzing the scripts, IP addresses of servers where malicious programs are hosted were collected, and all IP addresses were detected as malicious.

Along with the bash scripts, the attackers downloaded two malicious programs that were identified as ELF files. Collected malware samples were identified as botnet and cryptocurrency mining malware through reverse engineering and analysis. By analyzing their network traffic, contacted domains and IP addresses were identified, which were detected as malicious. Both malicious programs use the same way to maintain persistence, that is, they use the "cron" service to set scheduled tasks. The set task would be executed in certain time periods, and then the malicious program would be downloaded again. By reverse engineering the Sample2, the process of converting the executable file into a Python script was demonstrated, after the completion of the process a detailed insight into the functioning of the malware was obtained. In addition, another method of maintaining activity was detected in which the malicious program places its copies in directories that are activated every time the device is started. During the analysis, the tools Detect It Easy, Wireshark, Pyinstxtractor, Uncompyle6, Hatching Triage, and Joesandbox were used, as well as the built-in commands of the Linux operating system: "readelf" and "strings." The aforementioned methods and tools of reverse engineering proved to be an indispensable part of improving the security of information and communication systems. Based on knowledge about the functionalities of malicious programs obtained through reverse engineering, detection and prevention of the same or similar attacks is possible.

5.2 Protection Guidelines

Protection guidelines for improving the security of information and communication systems are based on the presented synthesis of results. Guidelines and their effects can be seen in Table 5.

These guidelines are based on increasing the security of remote connection services. Such services are used by attackers to gain initial, unauthorized access to a specific network or device; therefore, greater efforts are needed to increase their security. Other guidelines refer to the detection and prevention of active cases

Table 5 Protection guidelines and their effects

Guideline	Effect
Set limits on the number of authentication requests that a single IP address can generate	Decreases the possibility of successful brute-force attack
Strong passwords or authentication certificates need to be used if remote connection protocols are enabled for connecting to the server	Decreases the possibility of successful brute-force attack
Monitor connection attempts to servers to detect potential attacks	Based on the collected information, suspicious IP addresses can be blocked
Network traffic monitoring	If the attacker manages to gain access to the system, it is necessary to have the ability to detect suspicious network traffic that the attacker or malware generates through its activities. Such traffic can be detected if infected devices communicate with malicious IP addresses or domains. In such cases, TCP, HTTP, and DNS packets should be monitored and analyzed
File integrity monitoring	Through such solutions, it is possible to detect changes in file system that occur during the creation of new files, modification, or deletion of existing files
Use antivirus solutions	It is necessary to carry out regular scans of the entire system. Although antivirus solutions are mostly based on the recognition of already known activities of malicious programs, modern AV solutions have the ability to detect unknown malicious programs by analyzing processes
Regular security scans of network	Detect already known security vulnerabilities in the production network
Using honeypot servers and reverse engineering of malware	Organization can stay up-to-date with the latest attacks, which will then be easier to detect in the production network

of attacks, that is, those cases in which attackers successfully gain access to the system or network. In order to detect and prevent such attacks, it is necessary to actively monitor network traffic, suspicious file system activities, and suspicious modifications of active processes.

6 Conclusion

Statistical data indicate that we are in the midst of a cyberwar, where cybercriminals are the enemy and malware is the weapon. To guarantee the safety of all Internet users, it is necessary for researchers to keep pace with the development of malicious programs. The key to continued development of cybersecurity involves identifying and fixing security vulnerabilities, which is impossible without gathering adequate information about the characteristics and code of malicious programs. In this research, Cowrie was chosen because of its availability, configurability, data collection capabilities, and documentation. The implemented honeypot server collected two malware samples which were analyzed using reverse engineering methods and tools. Based on the performed analysis, the IP addresses and domains with which the malicious program comes into contact are shown, as well as the functionalities with which the collected malware maintains persistence within the system, such as cron. Analysis has shown that malware attacks delivered through poorly configured servers are common and executed using scripts to automate the whole process. To increase the security of servers and to increase malware detection rate, it is necessary to monitor network traffic and changes in the file system. By limiting the number of authentication requests and setting strong password requirements or only allowing connections based on certificates decreases the likelihood of successful brute-force or dictionary-based attacks. Honeypot servers and reverse engineering have proven to be invaluable tools to help identify malicious activity and improve the security of information and communication systems. Future research will focus on implementing honeypot servers hosted in different countries through chosen cloud provider to research the impact of geographical location on collected data such as used password and username lists, types of gathered malware and interactions with the server. This type of research will be used to paint a more complete picture of the malware landscape through which more detailed security guidelines can be proposed.

References

1. Statista, Malware - statistics & facts. https://www.statista.com/topics/8338/malware/#dossierKeyfigures. Accessed Dec 2022
2. Statista, Annual number of malware attacks worldwide from 2015 to first half 2022. https://www.statista.com/statistics/873097/malware-attacks-per-year-worldwide/. Accessed Dec 2022

3. I. Cvitić, D. Peraković, M. Periša, D. Sever, Defining cross-site scripting attack resilience guidelines based on BeEF framework simulation. Mobile Netw. Appl (2022). https://link.springer.com/article/10.1007/s11036-022-02052-z
4. N. Kambow, L.K. Passi, Honeypots: The need of network security. Int. J. Comput. Sci. Inf. Technol **5**(5), 6098–6101 (2014)
5. S. Yusirwan, Y. Prayudi, I. Riadi, Implementation of malware analysis using static and dynamic analysis method. Int. J. Comput. Appl **117**(5), 975 (2015)
6. A. Datta, K. Anil Kumar, D A., An emerging malware analysis techniques and tools: A comparative analysis. Int. J. Eng. Res. Technol **10**(4), 112 (2021)
7. L. Spitzner, The honeynet project: Trapping the hackers. IEEE Secur. Priv. **1**(2), 15–23 (2003)
8. M. Bombardieri, S. Castanò, F. Curcio, A. Furfaro, H.D. Karatza, Honeypot-powered malware reverse engineering, in *2016 IEEE International Conference on Cloud Engineering Workshop (IC2EW)*, (Berlin, Germany, 2016), pp. 65–69. https://doi.org/10.1109/IC2EW.2016.16
9. R. Sihwail, K. Omar, A. Zainol, A. Khairul, A survey on malware analysis techniques: Stat-ic, dynamic, hybrid and memory analysis. International Journal on Advanced Science Engineering and Information Technology. **8**, 4–2 (2018)
10. E. Gandotra, D. Bansal, S. Sofat, Malware analysis and classification: A survey. J. Inf. Secur. **05**, 56–64 (2014). https://doi.org/10.4236/jis.2014.52006
11. CCDCOE, *Malware Reverse Engineering Handbook*. https://ccdcoe.org/uploads/2020/07/Malware_Reverse_Engineering_Handbook.pdf. Accessed Dec 2022
12. V. Bhardwaj, V. Kukreja, C. Sharma, I. Kansal, R. Popali, *Reverse Engineering-A Method for Analyzing Malicious Code Behavior* (IEEE, 2022). https://doi.org/10.1109/ICAC353642.2021.9697150

A Method with Roulette Selection Strategy for Path Planning in UAV-Based Waste Monitoring Systems

Dalibor Dobrilovic, Jelena Stojanov, Dragan Peraković, Gordana Jotanovic, and Goran Jausevac

1 Introduction

The utilization of the Internet of Things (IoT) and relating new technologies, UAVs, robotics, and artificial intelligence, has a significant role in the efforts to make cities greener, safer, and more efficient. Connecting various devices, infrastructures, and vehicles can improve safety and quality of life in the entire city. According to United Nations research, by 2050, around 66% of the world's population will be living in cities [1]. The development of information and communication technologies provides possibilities and facilitates the implementation of artificial neural networks and modern communication technologies such as the Narrowband Internet of Things (NB-IoT). NB-IoT is a technology developed by 3GPP and represents Low Power Wide Area Network (LPWAN) radio technology standard. This technology is designed to enable a wide range of cellular services. Like other LPWAN technologies, NB-IoT focuses on low-cost outdoor and indoor coverage, with a high connection density, and long battery life. In favor of these reasons, this technology is very suitable for implementation in the system of control and management of waste in urban areas. Building such solutions, based on open standards, can be a serious

D. Dobrilovic (✉) · J. Stojanov
Technical Faculty "Mihajlo Pupin" Zrenjanin, University of Novi Sad, Zrenjanin, Serbia
e-mail: dalibor.dobrilovic@uns.ac.rs; jelena.stojanov@uns.ac.rs

D. Peraković
Faculty of Transport and Traffic Sciences, University of Zagreb, Zagreb, Croatia
e-mail: dperakovic@fpz.unizg.hr

G. Jotanovic · G. Jausevac
Faculty of Transport and Traffic Engineering, University of East Sarajevo, Doboj, Bosnia and Herzegovina
e-mail: gordana.jotanovic@sf.ues.rs.ba; goran.jausevac@sf.ues.rs.ba

L. Knapčíková, D. Peraković (eds.), *8th EAI International Conference on Management of Manufacturing Systems*, EAI/Springer Innovations in Communication and Computing, https://doi.org/10.1007/978-3-031-53161-3_6

challenge. The usage of other LPWAN technologies, such as LoRa and LoRaWAN, in combination with UAVs for smart city environments, is considered in [2].

The problem of waste management and waste collection is one of many problems in big cities. It is important to emphasize that currently almost all applications using the "Internet of Things" for smart waste management do not deal with household waste, but are focused on commercial waste and communal, public waste containers. Efficient waste collection is a necessary service in smart cities. The process of waste management can be significantly improved by the use of new technologies, such as UAVs. By integrating institutions, utility, and service companies from different fields, the best technological solutions can be achieved in smart cities and thus create a system that provides a new quality of life in urban and rural areas.

The traditional way of managing waste collection from waste containers has been to schedule and send large trucks around the city at regular intervals. When a truck arrives at the location of the container, it would be emptied regardless of whether the container was full or half empty. Some of the containers may be empty for a longer period while, depending on the structure and behavior of the citizens, some containers can be filled faster than usual, leading to excessive waste and garbage-related health hazards. The additional problem can be caused by uncontrolled garbage disposal in green areas such as parks, sports grounds, and picnic places. Therefore, these places should be monitored together with the public waste containers.

Another big problem is the illegal disposal of waste in places that are not planned for that purpose, resulting in unregulated landfills. Therefore, there is a need for constant monitoring of certain critical points to obtain timely information, providing vital data for further activities. To address all enlisted problems, this paper proposes the solution for the UAV-based system for monitoring, surveillance, and collection of data in critical areas.

This paper proposes a UAV-based IoT for observing and monitoring the presence of garbage with pre-calculated drone trajectories in urban environments. The contribution of this paper is the model with the proposed algorithm of random UAV path planning. The algorithm is based on Traveling Salesman Problem (TSP) with the Genetic Algorithm (GA) support and Roulette selection strategy for randomized selection of locations to visit. All locations are classified in three classes in accordance with the priority for monitoring, thus giving certain locations higher probability of visiting. This paper has the following structure. Section 2 gives an overview of related research in this domain. The proposed model of the system for monitoring and identifying garbage in public areas is presented in Sect. 3. Section 4 presents the random path planning algorithm of the proposed model in a specified environment based on TSP w. GA. Section 5 gives the examples of possible inclusion of VRP algorithms for path planning, with the addition of Roulette selection strategy to TSP solutions. In Sect. 6, the results of TSP w. GA and Roulette selection strategy are presented and discussed. The concluding remarks and a reference to future research are presented at the end of this paper.

2 Related Work

Currently, the problems of preserving green spaces in urban areas using UAVs are very important. One problem is the use of UAVs to observe and monitor the presence of garbage with the paths of UAVs in urban areas. The presented works deal with the use of UAV aircraft in urban areas in order to preserve the environment and improve the quality of human living environment.

According to [3], software-defined networking could facilitate flexible deployment and management of new services and help reduce costs, increase security and availability in networks. Network UAV routing requires protocols that would meet the conditions of high mobility, dynamic topologies, intermittent connections, power constraints, and variable quality connections.

UAVs equipped with micro wireless sensor nodes may contain applications such as environmental monitoring, disaster relief, military operations [4]. Situations of excessive aggregation and small amounts of dispersion of UAVs can occur during network formation. As a solution to the problem Fang et al. [4] propose a solution based on a 3D virtual force guided by a self-adaptive implementation (called 3DVFSD).

An interesting topic is the network communication architecture of UAV and vehicles in which UAVs fly over a certain area and provide communication services to the basic coverage area. In such cases, an ad hoc network for vehicles with the help of a UAV takes advantage of communication in line of visibility, load balancing, flexibility, and cost-effectiveness of scheduling [5]. Characteristics of the model proposed by Raza et al. [5] are evaluated against a highway collision case study. Model testing shows that the use of UAVs ensures guaranteed and timely delivery of emergency messages to nearby vehicles so that safe action can be taken and to avoid further damage.

UAVs are a rapidly evolving technology and, with their widespread use, can help detect illicit activities. UAVs can be sources of digital evidence, and more recently, their stored and transmitted data have been collected and analyzed. A paper published by Husnjak et al. [6] represents a forensic investigation of a UAV, specifically a DJI Mavic Air, and a related DJI GO 4 smartphone app installed on an Android device used as part of a GCS. The research was conducted in five steps and it was determined that the internal and external memory of DJI Mavic Air and the associated Android device with the DJI GO 4 application installed contain a significant and valuable amount of data that can be evidence in a digital investigation [6].

The aim of the research presented in [7] is to evaluate the efficiency of unmanned aerial vehicles in monitoring landfill settlements. The assessment is performed by comparing two models obtained by acquiring UAV images from two separate flights, repeated after approximately 6 months.

In smart cities, it is important to discover how collaborative drones and IoT improve the intelligence of smart cities [8]. This improvement can be achieved by collecting data and monitoring the quality of life while raising public safety, privacy, and security and optimizing energy consumption.

The system presented in [9] is an energy-efficient air quality monitoring system based on air and ground sensor. The architecture of this system consists of four layers: sensor layer, transmission layer, processing layer, and presentation layer. This implementation has been deployed at Beijing University and Xidian University since February 2018. Spatial adjustment and short-term forecasting are performed for data processing in order to eliminate the effects of incomplete measurement and delays in data loading.

IoT and UAV technology collaboration plays a vital role in green IoT. This collaboration enables a sustainable, reliable, environmentally friendly Industry 4.0. The research presented in [10] describes techniques and strategies that enable green IoT using drone infrastructure.

In [11], the possibilities of using drones to monitor garbage disposal at unauthorized locations are explored. The proposed system uses machine learning techniques and artificial intelligence to detect illicit debris in images collected by drones. The authors suggest that further research be based on the development of an automated robotic system for handling detected waste.

This paper presents a smart garbage detection system with image processing and the use of drones to capture images of garbage sites [12]. The authors use drones, with an Arduino microcontroller device, a deep neural network, the Python programming language, and Google's cloud platform.

Paper [13] considers the application of unmanned aerial vehicles in smart cities, examining their application as well as challenges. Considering the wide range of UAV applications in many areas, such as environmental hazards and pollution monitoring, and traffic management, the authors discussed challenges and issues such as safety, privacy, and ethical use of UAV.

An additional model for the model proposed in this paper and its further development can be taken from the experience in building an automated video surveillance system (AVS) designed to monitor objects in real time. One solution is a distributed video surveillance system based on a client-server architecture [14]. The system is accessible from portable devices such as tablets and smartphones. Also, systems can be focused on a realistic framework for simulating a wireless AVS system [15]. The framework models and simulates most of the details of a typical wireless AVS system. The proposed simulation framework is designed with the NS-3 network simulator.

3 The Model of the System

The model of UAVs-based urban waste monitoring system is presented in this section. The model is designed to monitor green urban areas, using the unmanned aerial vehicle (UAV). The role of UAV is to monitor the city region to prevent illegal waste dumps and containers overloads in public areas. Also, the goal of the model is to create a balanced low-cost and effective solution that will reduce the cost of the deployment and maintenance without reducing the efficiency of monitoring. The idea is to use a single or minimized number of UAVs to cover all specified locations. The model is built on the algorithm for UAV path planning based on the Traveling Salesman Problem (TSP) and Genetic Algorithm (GA). The model assumes the usage of the centralized system. Data processing is placed in the control center, but it is out of the focus of this model. The model is based on the usage of available and commercially popular drones. Furthermore, the equipment of commercial drones plays a significant role in shaping their price. The system has centralized data processing and drone management.

The system architecture is initially based on [16] and further developed in [17]. The system is modified with several vehicle routing algorithms and with proposed Roulette selection strategy. The addition of Roulette selection strategy allows dynamical selection of nodes with different priorities.

The proposed system consists of preferably one, but possibly more drones (depending on the size of the area to cover), and the control center. The key element of the system is surveillance drone for the detection of illegal garbage and unregulated garbage dumps on green urban areas. The supervising drone collects the images of defined green areas, areas with public garbage containers, and possible places for illegal garbage dumps. Collected images are uploaded to the server in the control center for further image processing and detection and identification of improperly disposed waste. Image processing software component is part of the control center, but it is not considered in this proposed model. The role of surveillance drones is to observe public areas. The drones should have support for route tracking to able to visit defined locations. The defined locations will be discussed later (Sect. 4), together with the algorithms used for drone path planning.

The full process is shown in Fig. 1. The first step (1) is path planning performed in path calculation module. The path definition and planning process is a central part of this paper and it is described in Sect. 4 and onward. After uploading a path, the supervising drone takes-off (2) and starts with the flight. The drone follows the calculated route (3) and supervises the locations along the route. After finishing the route, the drone (4) returns to control center and uploads images to the server's DB (6). Uploaded and stored images are later processed with the image processing module (7).

A Docking Station is considered as a fully automated system, with the support for operations such as managing drones, providing their vertical takeoff and landing, battery fast charging, and providing communication and data transfer between the control center and the drone. The Docking station has a wired network connection to the control center and the power supply infrastructure [15]. One of the most important components of the proposed model is the module for route calculation deployed in the control center. The process of path planning and route calculation is designed to enable the use of one or a minimal number of drones for covering the specified areas, thus reducing the UAV fleet costs and maintenance. First, the Traveling Salesman algorithm (TSP) with Genetic Algorithm (GA) is used to calculate the path for the UAV route. The calculated route is formed from the set of

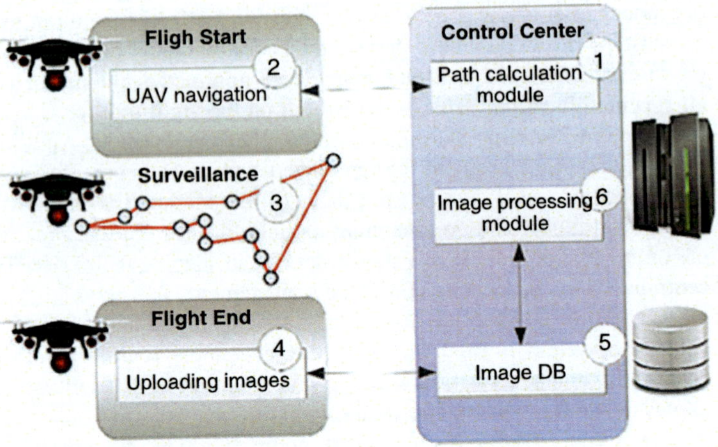

Fig. 1 System description and UAV operations algorithm

nodes of interest. The definition of the set of nodes of interest is based on Roulette selection system described in the following section. After the route calculation, the system checks if the route is longer than deployed UAV capability. If the route is longer, it is calculated again. If the route is within the UAV operational range, it is uploaded to the UAV.

4 The Path Definition

The evolution of the process of path calculation and the node set definition is described in this section. This process changed gradually. It started with the Traveling Salesman Problem (TSP) [18] solved with the Genetic Algorithm (GA) and random node selection. This first phase approach is later improved, first with the Vehicle Routing Problem (VRP) with three algorithms. Finally, the TSP with GA only approach is enhanced with the Roulette selection strategy. The later proposal is compared to the previous ones, and the comparison shows that the later proposal is efficient and applicable for surveillance and monitoring of urban waste disposal. The evolution process of path planning and node selection is described in the following text.

The process starts with the node set definition. This task is performed once at the start of system operation, for example, after system upgrade, or change of the waste management policy. The definition of the node set depends on the waste management policy, and therefore of targeted locations. In this paper, for example, the presented system is designed to monitor the unregulated waste disposal at the places of interest (such as public green areas). The system (drone) should take photos at the given locations, and forward collected photos for further processing.

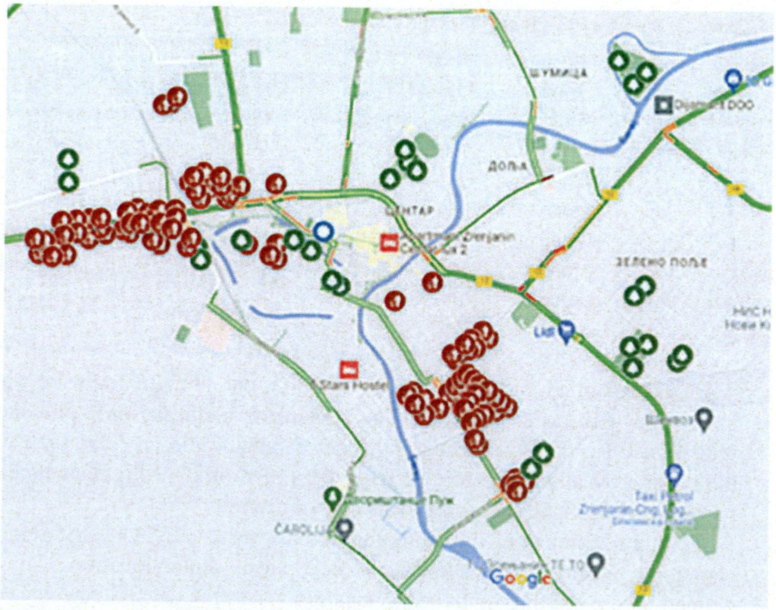

Fig. 2 The observing point locations in the city of Zrenjanin. (Source: Google maps)

The goal of processing photos is to search the green areas pollution incidents and unregulated waste detection. The defined node set is based on the need for monitoring the public places where people are gathering, such as parks, green surfaces, sports terrains, and bathing areas. The model of the system presented in this paper is designed to select a smaller random number of observing locations. The number of observing locations can be 10, 15, 20, or 30 as it was defined in these simulations. The selected number of locations is used to program the drone route each time before sending the drone to observation tasks. In this way, with the random location supervision which change at each turn, the possibility to monitor a large area with one drone is provided, causing a reduction in the investment in the system and its maintenance costs.

The example of the system operation is given for the city of Zrenjanin, a city in Serbia, located in the Central Banat region. The city is located in the northern part of Serbia (45°23′0//N, 20°23′22//E). The city urban area covers 193.03 km2 (74.53 sq mi) and the population is 76,511 (2011 census). The structure of observing points is 29 parks, green surfaces, sports, and bathing areas, 97 public garbage container locations, and one unregulated solid waste landfill (Fig. 2). The garbage container locations are not locations with a single container. To optimize the waste monitoring system the chosen container locations for monitoring are locations near multistory residential buildings where two or more containers are grouped. The system is modeled to monitor the over-filling of the containers, and garbage left

Table 1 Summary of simulation results

No.	Nodes	Iterations	Tests	Avg. (km)	Min (km)	Max (km)	St. Dev. (km)
1	10	1000	100	13.18	12.93	16.08	0.57
2	10	1000	500	13.26	13.14	15.12	0.19
3	15	1000	100	17.27	15.21	22.81	1.84
4	15	1000	500	15.75	14.25	21.72	1.22
5	20	1000	100	16.02	13.83	21.37	1.55
6	20	5000	100	16.38	14.94	21.52	1.38
7	30	1000	100	26.21	19.05	35.10	3.04
8	30	5000	100	26.83	22.42	35.96	2.56

near the over-filled containers. As an addition, one unregulated solid waste landfill is covered with these locations. The reason for including only one such location is that such locations are generally located outside the city. So, their inclusion in the coverage can significantly increase the range of UAV flights, which might be a problem considering the battery capacity of drones.

The total number of locations for monitoring is 127. One location in the city center is chosen for a headquarters and control center location. This location is defined as the starting and ending location for every calculated route, thus giving the total number of 128 locations. In this example, the Traveling Salesman algorithm (TSP) with Genetic Algorithm (GA) is used to calculate the path for 10, 15, 20, or 30 randomly selected observing locations [19]. The route calculation is simulated with Matlab/Octave code built upon the [19]. The code is modified to be applicable for the described scenario. The simulation uses changeable number of randomly selected nodes, number of iterations, and number of tests as well. The results are presented in Table 1.

The distance for the calculated path for each set of randomly selected nodes is calculated and recorded for the analyses. As one can see from Table 1, the average distances for 10 nodes are 10 km, for 15 and 20 nodes 15–17 km, and for 30 nodes around 26 km. The results also show that 1000 iterations give accurate results.

As an example, two routes with 10 and 15 randomly selected nodes are graphically presented in Fig. 3. The routes schedules are as follows: Center-L0014-L0019-L0020-L0022-L0119 -L0122 -L0132 -L0147 -L0165 -L0177-Center, and Center-L0004-L0108-L0137-L0149-L0115-L0133-L0140-L0191-L0102-L0188-L0160-L0170-L0182-L0180-L0157-Center. The results (Table 1) show that 10 or 15 locations can be easily monitored with the average route length of up to 15 km.

5 Further Model Expansion

In this section, the possible extensions of the model are presented. The first possible extension is based on Vehicle Routing Problem (VRP) with the utilization of

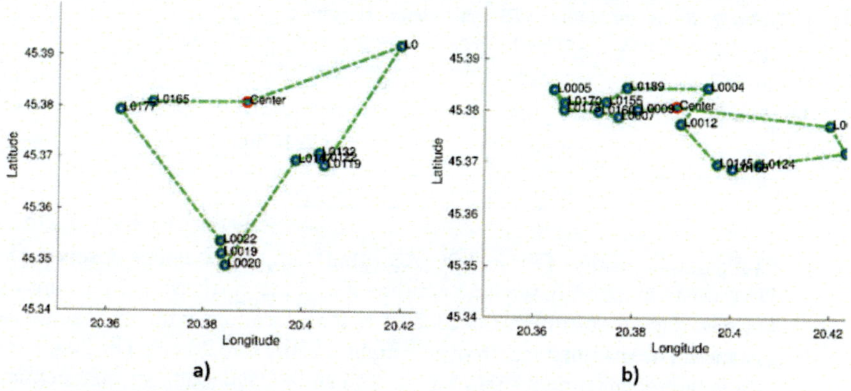

Fig. 3 The calculated routes with (**a**) 10 nodes, and (**b**) 15 randomly picked nodes

three different algorithms. In the second subsection, the statistics of location visits frequency is compared with VR visits statistic to investigate if locations will be monitored in a sufficient way, even in the case when the path is randomly planned. In the third section, the efficiency of implementation of Roulette selection strategy is analyzed, for the assumption that not all location should be equally monitored. For that purpose, the locations are divided into three classes.

5.1 VRP (Vehicle Routing Problem) Comparison

As an alternative for TSP solution, the VRP is considered [20–22]. There are several variants of VRP such as: Vehicle Routing Problem with travel distance constraint (DVRP), Vehicle Routing Problem with capacity constraint (CVRP), Vehicle Routing Problem with travel distance and capacity constraint (CDVRP), Vehicle Routing Problem with travel distance constraint, capacity constraint, and hard-time window constraint (VRPTW).

The usage of VRP solutions can be used for single drone or multiple drone scenarios equally. The calculated path can be used in scenario with the single drone, for making flight tours in sequences. In the case of multiple UAVs, the tours can be made simultaneously with the schedule in accordance with the drone and tours numbers.

The variant applicable for UAV path planning in urban waste monitoring scenarios is Vehicle Routing Problem with travel distance constraint (DVRP). The DVRP variant is implemented with three algorithms for comparison. Those three algorithms are Genetic Algorithm (GA), Simulated Annealing (SA), and Hybrid Particle Swarm Optimization with Genetic Algorithm involved (HPSO). The simulation is made again in Matlab/Octave code, and code is built upon [23].

Table 2 Summary of three VRP optimization algorithms

VRP Alg.	In. length	Opt. length	Iterations	Time	No. of Flights	Reduction
GA	302.34	217.39	100	126.97	17.22	28.10
HSPO	302.37	209.04	100	414.77	18.56	30.87
SA	342.88	114.33	132	179.31	14.64	66.66
Total					16.81	

Comparative results of these three algorithm's implementations in solving VRP for UAV waste monitoring scenario are shown in Table 2. All three algorithms calculate optimized number of flights needed to visit all locations with optimal number path distance. The maximal length of one flight is limited to 20 km. The 20 km is chosen as a value for maximum flight range, but can be changed to specific commercially available drone flight range.

The results in Table 2 are as follows: VRP Alg. – Type of algorithm used (GA, HSPO, and SA); In. length – initial cumulative path distance for all locations in km; Opt. length – optimized path distance for all locations in km; Iterations – number of iterations needed for calculating optimal distance; Time – time needed for calculating optimal distance; No. of Flights – minimal number of flights for calculated optimal distance; Reduction – reduction of optimized path length comparing to initial path length in percentage.

Basically, all three algorithms calculate initial path, optimized path length, time needed for calculation, best cumulative route, and route for each vehicle.

5.2 *TSP with Genetic Algorithm*

According to the simulation results presented in Table 2, where it is shown that three algorithms for VRP calculation give in average 16 flights needed to visit all locations (16.81). The TSP w. GA method is simulated and compared to VRP solutions. The frequency of visits per locations are calculated each 16 flights in five experiments with 500 runs of one experiment. The results are presented in Table 3.

The results presented in the Table 3 are as follows: FL – number of completed flights (rounds); AVG – average number of visits for all 127 locations; STD – standard deviation in number of visits for all 127 locations; No. Z – number of locations visited zero (0) times in the current round.

The distribution of number of visits per location after four cuts (64th, 192nd, 324th, and 452nd flight) are presented in Fig. 4. The x-axis presents the locations (1–127), and y-axis represents the number of visits. Figure 4a represents cumulative results for all five experiments, and Fig. 4b gives the number of visits of 1st experiment only.

The presented results show that after 17 rounds, the number of unvisited locations range from 11 to 24 (around 10–20%), and after 33 rounds, the number of unvisited

Table 3 Results of five experiments of TSP with GA optimization method

FL	1st experiment			2nd experiment			3rd experiment			4th experiment			5th experiment		
	AVG	STD	No. Z	AVG	STD	No. Z	AVG	STD	No. Z	AVG	STD	No. Z	AVG	STD	No. Z
1	0.12	0.32	112	0.12	0.32	112	0.12	0.32	112	0.12	0.32	112	0.12	0.32	112
17	2.01	1.47	24	2.01	1.37	17	2.01	1.19	11	2.01	1.49	22	2.01	1.43	18
33	3.90	1.89	0	3.90	1.93	1	3.90	1.68	2	3.90	1.94	3	3.90	1.89	3
49	5.79	2.45	0	5.79	2.41	0	5.79	2.05	1	5.79	2.53	0	5.79	2.42	1
65	7.68	2.82	0	7.68	2.73	0	7.68	2.58	0	7.68	2.74	0	7.68	2.75	0
81	9.57	3.28	0	9.57	3.06	0	9.57	2.92	0	9.57	3.04	0	9.57	3.06	0
97	11.46	3.30	0	11.46	3.12	0	11.46	3.14	0	11.46	3.44	0	11.46	3.39	0
113	13.35	3.36	0	13.35	3.29	0	13.35	3.31	0	13.35	3.67	0	13.35	3.67	0
129	15.24	3.69	0	15.24	3.49	0	15.24	3.47	0	15.24	3.86	0	15.24	3.93	0
145	17.13	3.95	0	17.13	3.69	0	17.13	3.84	0	17.13	3.89	0	17.13	4.36	0
161	19.02	4.16	0	19.02	3.96	0	19.02	4.17	0	19.02	4.06	0	19.02	4.70	0
177	20.91	4.43	0	20.91	4.11	0	20.91	4.49	0	20.91	4.23	0	20.91	4.91	0
193	22.80	4.47	0	22.80	4.20	0	22.80	4.66	0	22.80	4.31	0	22.80	5.33	0
209	24.69	4.76	0	24.69	4.52	0	24.69	4.61	0	24.69	4.39	0	24.69	5.26	0
225	26.57	4.89	0	26.57	4.88	0	26.57	4.84	0	26.57	4.49	0	26.57	5.53	0

(continued)

Table 3 (continued)

	1st experiment			2nd experiment			3rd experiment			4th experiment			5th experiment		
FL	AVG	STD	No. Z	AVG	STD	No. Z	AVG	STD	No. Z	AVG	STD	No. Z	AVG	STD	No. Z
241	28.46	5.07	0	28.46	4.99	0	28.46	4.95	0	28.46	4.58	0	28.46	5.85	0
257	30.35	5.13	0	30.35	5.12	0	30.35	5.17	0	30.35	4.77	0	30.35	6.00	0
273	32.24	5.37	0	32.24	5.12	0	32.24	5.34	0	32.24	4.94	0	32.24	6.03	0
289	34.13	5.63	0	34.13	5.42	0	34.13	5.41	0	34.13	5.23	0	34.13	6.28	0
305	36.02	5.71	0	36.02	5.55	0	36.02	5.56	0	36.02	5.33	0	36.02	6.48	0
321	37.91	5.99	0	37.91	5.85	0	37.91	5.43	0	37.91	5.60	0	37.91	6.32	0
337	39.80	6.06	0	39.80	6.00	0	39.80	5.61	0	39.80	5.72	0	39.80	6.59	0
353	41.69	6.22	0	41.69	6.08	0	41.69	5.96	0	41.69	5.98	0	41.69	6.74	0
369	43.58	6.24	0	43.58	5.89	0	43.58	6.18	0	43.58	6.07	0	43.58	6.73	0
385	45.47	6.51	0	45.47	6.04	0	45.47	6.33	0	45.47	6.32	0	45.47	6.88	0
401	47.36	6.50	0	47.36	6.12	0	47.36	6.50	0	47.36	6.48	0	47.36	7.16	0
417	49.25	6.75	0	49.25	6.27	0	49.25	6.58	0	49.25	6.67	0	49.25	7.17	0
433	51.14	6.88	0	51.14	6.63	0	51.14	6.67	0	51.14	6.92	0	51.14	7.35	0
449	53.03	7.17	0	53.03	6.68	0	53.03	6.88	0	53.03	7.24	0	53.03	7.55	0
465	54.92	7.43	0	54.92	6.87	0	54.92	6.93	0	54.92	7.59	0	54.92	7.32	0
481	56.81	7.75	0	56.81	6.97	0	56.81	7.08	0	56.81	7.76	0	56.81	7.32	0
497	58.70	7.86	0	58.70	7.13	0	58.70	7.44	0	58.70	7.88	0	58.70	7.37	0

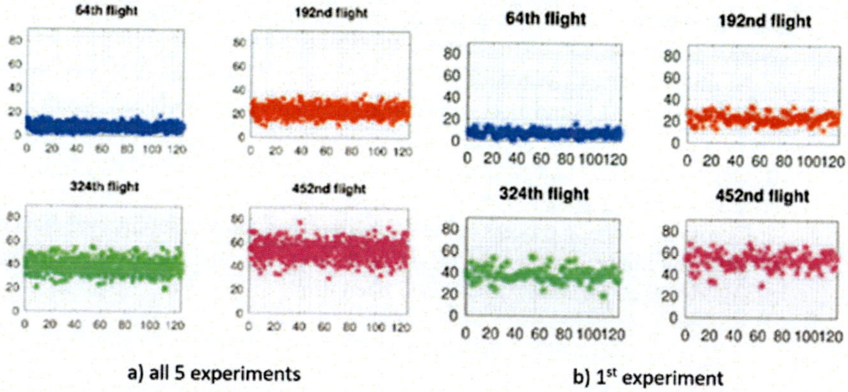

Fig. 4 Distribution of visits per location. (**a**) all five experiments. (**b**) 1st experiment

locations range from 0 to 5 (up to 4%), in all five experiments. Comparing with the VRP where on average 16 rounds should be taken to visit all nodes, TSP w. GA method with randomly selected location prove to be equally applicable.

According to the results, all locations will be monitored in a sufficient way, even in the case when the path is randomly planned. The randomly planned location visiting and reconfiguration of the path routes has an advantage, especially in monitoring scenarios. In this way, the UAV route could not be anticipated by the persons who intentionally are going to violate the waste disposal and environmental regulations, with unregulated garbage disposal.

5.3 *The Roulette Selection Strategy*

Starting with the assumption that not all locations should be equally monitored, the plan was to expand the model with random location selection mechanism. For this goal, the implementation of Roulette selection strategy is considered in this model.

Roulette selection is based on fitness proportionate selection [24], and it is commonly used within evolutionary algorithms, especially within Genetic Algorithm where the roulette wheel method determines the selection operator [25].

The real problem of choosing 15 locations out of 127 demands fitness values assigned to the locations based on the necessity to supervise some of them more often. Therefore, the locations are not equable in the node selection process.

The fitness values are non-uniformly distributed. Their determination methodology could be based on social, objective, empirical (or some others) facts, and it is not in the focus of this paper. Important property of the fitness values is that their sum is equal to one, so they represent a chance of location to be selected. In this paper, the fitness values will be used to distinguish the locations in three classes of various priorities.

A fitness values vector, $f = (f1, f2, \ldots, f127)$ uniquely determines the cumulative fitness values vector cf whose components cf1 = f1, cfk = cfk − 1 + fk, for $k = 2,\ldots,127$, can be seen as separating points of the unit interval (0,1). The formed subintervals unambiguously correspond to the locations, and their wideness reflects the fitness values.

The node selection process is based on identification of a subinterval that contains a random number between 0 and 1. Each iteration also includes checking if the subinterval (i.e., location) has already been used.

6 Results

The model is expanded with Roulette selection strategy. The simulation is designed and tailored to compare the model with previously presented results. In Table 4 are presented results of three experiments, each experiment consisting of 500 runs. In each experiment, 127 nodes are classified into three classes. The classes A, B, and C represent the priority of certain locations for monitoring. The 25 locations (20%) are classified as Class A. The fitness value of all locations in Class A is 0.013953488. It is three time higher than basic fitness value. The value of basic priority is 0.004651163. Furthermore, 38 locations (30%) are classified as Class B with fitness value 0.009302326 (two times higher than basic fitness value), and 64 locations (50%) are classified as Class C with basic fitness value (0.004651163).

The results presented in Table 4 are as follows: FL – number of completed flights; AVGA – average number of visits for first 20% nodes (Class A priority); STDA – standard deviation in number of visits for first 20% nodes (Class A priority); AVGB – average number of visits for second 30% nodes (Class B priority); STDB – standard deviation in number of visits for second 30% nodes (Class B priority); AVGC – average number of visits for third 50% nodes (Class C priority); STDC – standard deviation in number of visits for third 50% nodes (Class C priority); No. Z – number of locations visited zero (0) times.

The distribution of number of visits per location after four cuts (64th, 192nd, 324th, and 452nd flight) for model expanded with Roulette selection strategy are presented in Fig. 5. Figure 5a represents cumulative results for all three experiments, and Fig. 5b gives the number of visits in 1st experiment only.

The expansion of the model with Roulette selection strategy proved to be justified with these results. The results presented in Table 4 show that after 81 flights, there will be no unvisited locations. Furthermore, in Table 4, after 497 flights, the average number of visits for Class A is 102.45, for Class B is 62.09, and for Class C is 38.72 per locations in three experiments cumulatively. According to the results, it may be concluded that using proposed method, all locations will be sufficiently monitored, with a higher frequency of visits for defined most important or prioritized locations.

Table 4 Results of TSP/GA with Roulette selection strategy

	1st experiment							2nd experiment							3rd experiment						
FL	AVGA	STDA	AVGB	STDB	AVGC	STDC	No. Z	AVGA	STDA	AVGB	STDB	AVGC	STDC	No. Z	AVGA	STDA	AVGB	STDB	AVGC	STDC	No. Z
1	0.27	0.46	0.15	0.37	0.03	0.18	112	0.19	0.41	0.15	0.37	0.06	0.25	112	0.23	0.41	0.13	0.34	0.08	0.27	112
17	3.62	2.05	2.03	1.40	1.32	1.24	25	3.35	1.73	1.97	1.56	1.48	1.13	20	3.81	1.96	2.08	1.66	1.27	0.94	22
33	6.81	2.78	3.77	2.25	2.73	1.60	5	7.12	2.11	3.90	2.25	2.57	1.59	7	6.96	3.15	4.31	2.50	2.41	1.33	4
49	9.96	2.97	5.85	2.84	4.00	1.89	2	10.27	2.97	5.87	2.82	3.87	1.84	3	10.27	3.60	6.38	3.21	3.60	1.76	1
65	13.50	3.90	7.74	3.43	5.22	2.28	2	13.04	3.55	8.03	3.48	5.24	1.87	0	13.65	3.87	8.28	3.80	4.87	1.96	0
81	16.88	4.34	9.85	4.22	6.37	2.59	0	16.62	4.04	10.26	4.33	6.25	2.33	0	17.00	4.80	10.21	4.28	6.14	2.38	0
97	19.88	5.37	11.82	4.95	7.73	3.12	0	19.54	4.32	12.18	4.70	7.73	2.74	0	20.50	5.12	12.15	5.02	7.37	2.45	0
113	23.38	5.86	13.72	4.94	8.95	3.34	0	22.85	4.70	14.21	5.58	8.94	2.90	0	23.92	5.29	14.26	5.58	8.51	2.60	0
129	27.19	6.88	15.62	5.56	10.05	3.37	0	26.15	5.33	16.26	6.08	10.17	3.19	0	26.81	5.54	16.28	6.11	9.89	2.74	0
145	30.65	8.31	17.49	6.11	11.32	3.65	0	29.65	5.88	18.13	6.76	11.49	3.32	0	30.12	6.60	18.38	6.98	11.06	2.99	0
161	33.73	8.67	19.51	7.06	12.65	3.85	0	32.96	5.68	19.92	7.63	12.87	3.45	0	33.46	7.45	20.69	7.76	12.10	3.24	0
177	36.46	8.98	21.79	7.34	13.97	4.10	0	36.04	6.09	22.36	8.24	13.95	3.67	0	36.92	8.03	22.59	8.41	13.35	3.40	0
193	39.92	9.25	23.85	7.99	15.11	4.30	0	39.00	6.41	24.41	8.91	15.30	3.91	0	40.00	9.05	24.74	9.12	14.59	3.64	0
209	43.69	9.55	25.90	8.79	16.13	4.36	0	42.35	7.05	26.69	9.85	16.38	4.13	0	43.50	9.18	26.77	9.57	15.79	4.09	0
225	46.85	9.96	27.85	9.71	17.51	4.36	0	46.00	7.90	28.74	10.30	17.46	4.20	0	46.81	10.14	28.62	10.35	17.11	4.27	0
241	49.92	10.62	29.74	10.29	18.89	4.52	0	49.46	7.93	30.49	10.86	18.83	4.48	0	49.62	10.43	30.77	11.35	18.49	4.52	0

(continued)

Table 4 (continued)

	1st experiment							2nd experiment							3rd experiment						
FL	AVGA	STDA	AVGB	STDB	AVGC	STDC	No. Z	AVGA	STDA	AVGB	STDB	AVGC	STDC	No. Z	AVGA	STDA	AVGB	STDB	AVGC	STDC	No. Z
257	53.46	10.79	31.56	10.76	20.11	4.76	0	52.77	8.54	32.77	11.46	19.92	4.49	0	53.04	10.90	32.62	11.84	19.78	4.51	0
273	56.38	11.60	33.46	10.92	21.56	4.95	0	56.12	8.34	34.95	12.08	21.02	4.82	0	56.23	11.23	34.77	12.18	20.97	4.55	0
289	59.92	12.19	35.08	11.39	22.94	5.03	0	60.04	8.29	36.56	12.36	22.21	5.15	0	59.42	11.59	37.00	12.70	22.10	4.51	0
305	63.50	12.39	36.95	12.06	24.14	5.31	0	63.23	9.05	38.72	12.89	23.40	5.15	0	62.85	12.79	38.77	13.25	23.44	4.63	0
321	66.85	12.44	38.90	12.44	25.40	5.35	0	66.31	9.27	40.87	13.41	24.62	5.11	0	66.08	12.90	40.87	13.38	24.63	4.81	0
337	70.23	13.00	40.82	13.30	26.63	5.25	0	70.31	9.95	42.64	14.12	25.71	5.40	0	69.69	13.12	42.97	14.06	25.68	4.97	0
353	73.35	13.80	42.82	13.79	27.92	5.45	0	73.12	10.22	44.87	14.71	27.00	5.28	0	72.69	13.70	45.21	14.93	26.90	4.98	0
369	76.31	13.99	44.82	14.30	29.29	5.60	0	76.73	11.30	46.82	15.55	28.19	5.44	0	76.15	13.57	47.18	15.30	28.08	5.16	0
385	79.08	14.46	46.92	14.96	30.67	5.72	0	80.00	11.77	48.46	16.07	29.65	5.71	0	79.69	13.32	49.15	15.77	29.25	5.14	0
401	81.81	15.16	49.15	15.40	32.03	5.86	0	83.27	12.54	50.44	16.71	30.89	5.81	0	82.88	13.80	51.26	16.46	30.48	5.24	0
417	84.85	15.53	51.59	15.69	33.11	5.97	0	86.65	12.96	52.38	17.23	32.16	6.01	0	86.15	13.79	52.95	17.08	31.94	5.46	0
433	88.00	16.20	53.49	16.32	34.48	6.18	0	90.00	13.28	54.46	17.80	33.33	6.20	0	89.69	14.24	54.82	17.55	33.17	5.43	0
449	91.69	16.76	55.54	17.04	35.56	6.46	0	93.19	13.63	56.28	18.40	34.70	6.79	0	92.88	14.83	56.56	18.27	34.60	5.76	0
465	94.65	16.36	57.51	17.38	36.95	6.48	0	96.69	14.23	58.31	18.78	35.83	6.74	0	96.31	15.36	58.56	18.68	35.81	5.76	0
481	98.08	16.73	59.18	17.86	38.35	6.47	0	100.46	14.73	60.18	18.94	36.92	6.88	0	99.88	15.62	60.54	19.27	36.98	5.75	0
497	101.00	17.28	61.26	18.31	39.67	6.88	0	103.50	14.71	62.26	19.25	38.25	6.94	0	102.85	16.26	62.77	19.61	38.24	5.96	0

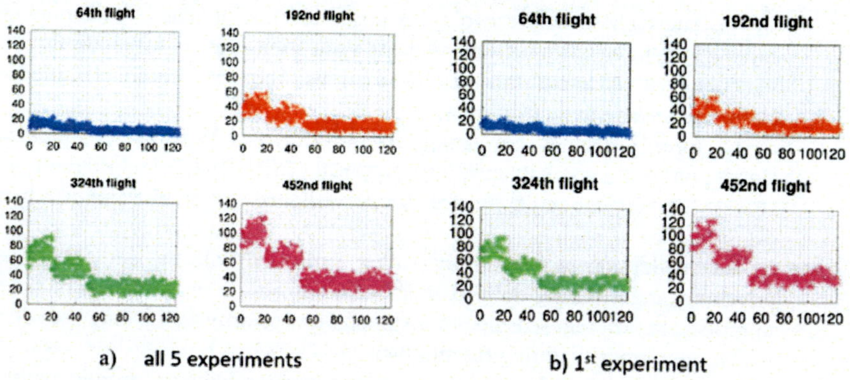

Fig. 5 Distribution of visits per location. (**a**) all five experiments. (**b**) 1st experiment

7 Conclusion

The concern for monitoring urban public green areas and waste disposal locations with the utilization of drones is the focus of this paper. Urban environments' public green areas, such as parks, green surfaces, sports terrains, and bathing areas, are gathering places for people and therefore prone to unregulated waste disposal. Other places with similar concerns related to garbage disposal violations are locations where the group of public garbage containers are positioned. Those locations are near multistory residential buildings, where a large number of people use those containers. As a potential solution for a problem, an approach of using UAVs for urban waste monitoring is presented. The proposed solution is modeled to monitor green areas, public trash cans, garbage containers, and unregulated landfills, a total 127 locations. The goal of this monitoring is to detect possible violations of garbage disposal rules.

This solution is focused on a method for random path planning and location selection, allowing utilization of a minimal number of UAVs. The idea of the proposed system is to select a random path for monitoring smaller but changeable set of locations in each iteration of drone usage. This will lead to the balance of drone utilization, aerial coverage, and monitoring efficiency, and will give the opportunity for the lower number of drones to efficiently monitor larger areas. This approach will lead to lower system purchase and maintenance costs.

This paper has presented several path planning approaches. The idea evolved at the start with the usage of Genetic Algorithm (GA) for Traveling Salesman Problem (TSP) solving. For this approach, the random set of smaller number of nodes with equal probabilities is chosen. The number of 10, 15, and 20 nodes are investigated.

The second phase tries to investigate the usage of three Vehicle Routing Problem (VRP) algorithms, respectively, with Genetic Algorithm (GA), Simulated Annealing (SA), and Hybrid Particle Swarm Optimization with Genetic Algorithm involved (HPSO).

The latest phase in the evolution was the return to the GA support for TSP solving with three introduction of Roulette selection strategy for reduced set of node selections with different priorities for locations. The locations are classified in three classes according to their priority.

All three proposals are simulated in Octave software and compared. The simulation results show that the later proposal (TSP with GA and Roulette selection strategy) is efficient and applicable for surveillance and monitoring of urban waste disposal.

The model can be potentially further expanded with the addition of location priority changes during defined time period, enabling the adoption of the system to be used in different seasons, for example, low priority for bathing area during the winter, high priority during the summer.

Furthermore, the addition of algorithms such as Dijkstra shortest path, A-star, D-star, Artificial Bee Colony (ABC), and Artificial Neural Networks (ANN) and others for planning routes between two locations for avoiding the urban obstacles will be one of the possible development directions.

References

1. K. Pardini, J.J.P.C. Rodrigues, S.A. Kozlov, N. Kumar, V. Furtado, IoT-based solid waste management solutions: A survey. J. Sens. Actuator Netw. **8** (2019). https://doi.org/10.3390/jsan8010005
2. L.-Y. Chen, H.-S. Huang, C.-J. Wu, Y.-T. Tsai, Y.-S. Chang, A LoRa-based air quality monitor on unmanned aerial vehicle for smart city. Presented at the 2018 international conference on system science and engineering (ICSSE), 2018
3. L. Gupta, R. Jain, G. Vaszkun, Survey of important issues in UAV communication networks. IEEE Commun Surv Tutor **18**, 1123–1152 (2015)
4. Y. Fang, B. Cheng, K. Kang, H. Tan, A self-adaptive deployment model of UAV cluster for emergency communication network. Int. J. Distrib Sensor Netw **17**, 15501477211049328 (2021)
5. A. Raza, S.H.R. Bukhari, F. Aadil, Z. Iqbal, An UAV-assisted VANET architecture for intelligent transportation system in smart cities. Int. J. Distrib. Sensor Netw **17**, 15501477211 (2021)
6. S. Husnjak, I. Forenbacher, D. Peraković, I. Cvitić, UAV forensics: DJI Mavic air noninvasive data extraction and analysis. Presented at the 5th EAI international conference on management of manufacturing systems, 2022
7. V. Baiocchi, Q. Napoleoni, M. Tesei, G. Servodio, M. Alicandro, D. Costantino, UAV for monitoring the settlement of a landfill. Eur. J. Remote Sens **52**, 41–52 (2019)
8. S.H. Alsamhi, O. Ma, M.S. Ansari, F.A. Almalki, Survey on collaborative smart drones and internet of things for improving smartness of smart cities. IEEE Access. **7**, 128125–128152 (2019)
9. Z. Hu, Z. Bai, Y. Yang, Z. Zheng, K. Bian, L. Song, UAV aided aerial-ground IoT for air quality sensing in smart city: Architecture, technologies, and implementation. IEEE Netw. **33**, 14–22 (2019)
10. S.H. Alsamhi, F. Afghah, R. Sahal, A. Hawbani, M.A. Al-qaness, B. Lee, M. Guizani, Green internet of things using UAVs in B5G networks: A review of applications and strategies. Ad Hoc Netw. **117**, 102505 (2021)

11. A.P. Anadkat, B. Monisha, M. Puthineedi, A.K. Patnaik, R. Shekhar, R. Syed, Drone based solid waste detection using deep learning & image processing. Presented at the alliance international conference on artificial intelligence and machine learning (AICAAM), 2019
12. P. Achaliya, G. Bidgar, H. Bhosale, P. Dhole, K. Gholap, S. Chandwad, Drone based smart garbage monitoring system using computer vision. Int. J. Creat. Res. Thoughts **8**, 1066 (2020)
13. F. Mohammed, A. Idries, N. Mohamed, J. Al-Jaroodi, I. Jawhar, UAVs for smart cities: Opportunities and challenges. Presented at the 2014 international conference on unmanned aircraft systems (ICUAS), 2014
14. P. Albano, A. Bruno, B. Carpentieri, A. Castiglione, A. Castiglione, F. Palmieri, R. Pizzolante, I. You, A secure distributed video surveillance system based on portable devices. CD-ARES **7465**, 403–415 (2012)
15. M.A. Alsmirat, Y. Jararweh, I. Obaidat, B.B. Gupta, Automated wireless video surveillance: An evaluation framework. J. Real-Time Image Proc. **13**, 527–546 (2017)
16. G. Jausevac, D. Dobrilovic, V. Brtka, G. Jotanovic, D. Perakovic, Z. Stojanov, Smart UAV monitoring system for parking supervision. Presented at the future access enablers for ubiquitous and intelligent infrastructures: 5th EAI international conference, FABULOUS 2021, Virtual Event, May 6–7, 2021, Proceedings, 2021
17. D. Dobrilovic, G. Jotanovic, A. Stjepanovic, G. Jausevac, D. Perakovic, A model of UAV-based waste monitoring system for urban areas, in *Cyber Security, Privacy and Networking: Proceedings of ICSPN 2021*, (Springer, 2022), pp. 309–319
18. W.J. Cook, *In Pursuit of the Traveling Salesman* (Princeton University Press, 2012). https://doi.org/10.1515/9781400839599
19. V. Chernov, Solution of Traveling Salesman Problem (TSP) using Genetic Algorithm (GA) in Matlab/Octave, https://github.com/vchernov/tspm (2016)
20. G.B. Dantzig, J.H. Ramser, The truck dispatching problem. Manag. Sci. **6**, 80–91 (1959). https://doi.org/10.1287/mnsc.6.1.80
21. G. Clarke, J.W. Wright, Scheduling of vehicles from a central depot to a number of delivery points. Oper. Res. **12**, 568–581 (1964). https://doi.org/10.1287/opre.12.4.568
22. P. Toth, *The Vehicle Routing Problem* (SIAM, 2002)
23. A MATLAB, https://github.com/liukewia/Solving-TSP-VRP/ (2021)
24. T. Back, *Evolutionary Algorithms in Theory and Practice: Evolution Strategies, Evolutionary Programming, Genetic Algorithms* (Oxford University Press, 1996)
25. P. Sharma, A. Wadhwa, K. Komal, Analysis of selection schemes for solving an optimization problem in genetic algorithm. Int. J. Comput. Appl **93**, 1–3 (2014). https://doi.org/10.5120/16256-5714

Analyzing the Efficiency of Software Distribution Systems in AIoT Environments

Aleksandar Jevremovic, Mladen Veinovic, and Dragan Peraković

1 Introduction

In the past few decades, the capabilities of hardware platforms designed for the development of specialized devices (such as mobile phones or embedded computers) have advanced drastically. One significant consequence of this evolution is the opening up of possibilities for using standard software (e.g., the Linux operating system, standard libraries, etc.) on these platforms [1]. In most scenarios, it is no longer necessary to develop specialized software that fits within the constraints imposed by the hardware. As a result, we have obtained smartphones or IoT devices that use the same codebase as desktop computers. Or, it is possible to replace primitive microcontrollers that use specialized languages and development environments, with kernels that provide standardized system calls, implement the entire TCP/IP stack, and support a wide range of hardware devices.

The latest phase of this evolution is represented by the emergence of AIoT (Artificial Intelligence of Things) devices, specifically IoT devices that have sufficient hardware resources to execute complex and demanding artificial intelligence algorithms locally. From a hardware perspective, this involves having a fast enough CPU supplemented with a GPU/TPU for specific types of tasks, as well as an adequate amount of memory to store libraries, algorithms, data, and results. From a software perspective, artificial intelligence often involves the use of large frameworks written in high-level programming or pseudo-languages (such

A. Jevremovic (✉) · M. Veinovic
Faculty of Informatics and Computing, Singidunum University, Belgrade, Serbia
e-mail: ajevremovic@singidunum.ac.rs; mveinovic@singidunum.ac.rs

D. Peraković
Faculty of Transport and Traffic Sciences, University of Zagreb, Zagreb, Croatia
e-mail: dragan.perakovic@fpz.hr

L. Knapčíková, D. Peraković (eds.), *8th EAI International Conference on Management of Manufacturing Systems*, EAI/Springer Innovations in Communication and Computing, https://doi.org/10.1007/978-3-031-53161-3_7

as Python). Given the intensive development of artificial intelligence today, the focus is primarily on achieving better accuracy and effectiveness [2]. Performance considerations usually take a backseat, and questions about code optimization and solution distribution often do not arise during the solution's lifespan.

Performance and algorithm optimization are less emphasized in cloud environments. The main reasons for this are the abundance of resources and a lesser emphasis on energy efficiency. On the other hand, in (A)IoT environments, especially in cases where devices are battery-powered and their autonomy is crucial, performance and algorithm optimization can often be more significant criteria than result accuracy and precision. In fact, specialized chips with reduced precision but improved performance and energy efficiency are being developed today [3].

Until recently, there has not been much focus on optimizing the distribution of software solutions. Traditional package distribution and maintenance systems embedded in the operating system (e.g., YUM/DNF, APT, Slackpkg, etc.) or developed based on standard architectures for specific programming languages or software sets (e.g., Anaconda, PIP, npm, CPAN, PEAR, etc.) have been commonly used [4]. However, due to several factors, such as a shift in software development philosophy toward more frequent and smaller updates, limited resources in IoT environments, the need for automated updates, increasingly bulky software packages, etc., the question of efficient software distribution and related updates is becoming increasingly important [5].

In this chapter, we provide an overview of standard solutions and approaches for software package distribution and management, as well as an analysis of their efficiency. We consider classical package management systems, package distribution systems for specific programming languages, and hybrid systems of this kind. The analysis primarily focuses on IoT systems characterized by the need for efficient utilization of local resources and communication channels. Considering the widespread use of Linux in IoT devices, the analysis is tied to that operating system.

2 Overview of Traditional Software Distribution Approaches

2.1 Source Code Distribution

Downloading the source code, usually directly from the author's website, and compiling it on the device where it is intended to be executed is a basic and traditional method of open-source software distribution. This approach offers significant advantages compared to others, but it also has notable drawbacks. Nowadays, there are advanced software tools that automate the process of downloading the source code and compiling it (e.g., Pacman on Arch Linux or (Mac)Ports on MacOS/FreeBSD/OpenBSD systems).

One of the most important advantages of this approach to software is the ability to inspect and modify the downloaded code. Such capability is critically important for systems that handle important and sensitive data, as the use of closed-source software could potentially lead to the execution of undocumented/hidden/malicious functions. Many Linux distributions, for this reason, do not include any software package whose source code is not available, such as closed drivers, codecs, etc. However, it should be noted that inspecting source code, primarily to identify intentionally inserted backdoors, is not a simple, easy, or often practical task [6]. For example, the Linux source code, version 5.15, contains over thirty million lines of code, which would require significant human resources to thoroughly review.

Modifying the downloaded code before its usage is one of the exclusive advantages of free software. A simpler form of code modification involves altering its configuration before compilation. This entails enabling/disabling specific components/functions or setting certain parameters before generating the executable code. For the automation of the compilation process, such as for automated updates upon the release of new software versions, the configure/build script can be reused. However, if the changes in the new version are not backward-compatible, the automated compilation process may fail or result in incorrect outcomes.

Another form of code modification is the specific alteration of original functionalities defined in the code. Such modifications can be useful in situations where long-term support for a particular system is not necessary or when improving the modified software package versions. However, this approach also brings along all the problems and costs associated with forking.

Another important advantage of compiling source code on the target system is the ability to optimize performance and resource utilization. By compiling for a specific platform and instruction set, greater efficiency can be achieved during execution. Additionally, by excluding unused components/functions and optimizing the configuration of included components, the amount of memory utilized by the software during execution can be reduced.

Using source code for software distribution, however, has significant drawbacks, which is why it is rarely practiced in production environments. The compilation process itself, depending on the size and complexity of the software, can require substantial time and processing resources. This can be problematic for systems that operate at high utilization levels (e.g., data center servers, embedded systems, etc.) or for battery-powered systems. Additionally, there are often situations where a larger number of computers actually represent multiple instances with identical hardware–software configurations. In such cases, compiling the source code on each system would introduce unnecessary redundancy and resource wastage.

A significant drawback of compiling source code on production systems is the uncertainty and unreliability of the process due to its complexity. There are multiple reasons why the compilation process may fail, such as the absence of required components, lack of resources, incompatibility with the used build script, etc.

Another disadvantage of using software packages through source code is the need for additional tracking and record-keeping at the system level regarding installed software packages and their associated files. Such record-keeping becomes necessary when removing specific software packages or replacing their files with a new version. Some software packages have well-defined rules for uninstallation, which can be accomplished, for example, with the "make uninstall" command. However, this is not always the case, and it also assumes the preservation of the source code on the system where the software is installed.

Due to the mentioned drawbacks in production environments, the use of source code as a form of accessing a specific software package is rarely utilized. The mentioned advantages can be significant in certain scenarios, but this is typically not the case in production environments where the priorities are simple and automated administration, efficient resource utilization, and high stability and availability. Of course, for solutions based on interpreted programming languages, this form of software distribution is somewhat implicit, but alternative methods are often used even for them due to their own drawbacks.

2.2 Version Control Systems and Continuous Integration

A more granular approach to distributing source code includes version control systems. These systems offer advanced functionalities such as change tracking, multiple branch development, team collaboration, etc. However, these systems are primarily used in development and testing environments. In production environments, they are most commonly used within continuous integration. However, the specificities of IoT environments usually do not make this type of software distribution optimal.

2.3 Package Managers

The most popular way of handling software packages in Linux distributions (and therefore in IoT environments) is by using package managers. Package managers are a special type of system software whose primary task is to install, update, and remove software packages on an individual computer system. Package managers are usually specific to distributions—for example, YUM/DNF for RedHat distributions, APT for Debian-based distributions, pkgtools/slackpkg for Slackware distributions, etc. Some package managers, such as Flatpak [7], Snap [8], and similar ones, are not tied to a specific distribution.

They usually have the capability to automatically download packages from network repositories, verify integrity and authenticity, as well as determine and handle dependencies. The use of digital signatures and integrity checks allows packages to be downloaded from alternative repositories. In IoT environments, it is common practice to use repositories specific to IoT solution providers and to host servers near the production environment to facilitate faster package downloads and save bandwidth to the Internet.

Package managers often have built-in support for managing dependencies. This means that the manager itself can determine whether all the packages that the package being installed/updated depends on are installed on the system. If not, the manager will fetch and install the required packages. This dependency check is performed recursively. There are also situations where installing one package requires the installation of several or even dozens of additional packages due to dependencies.

There are several issues related to package dependencies. The previously mentioned situation of chain or immediate dependencies can result in the installation of a large number of packages. This leads to a significant consumption of space on external memory, communication channels, and CPU time for package downloading and installation/updates, as well as the time required to complete the process. However, more complex problems arise from conflicting and cyclical dependencies. In such situations, different packages depend on different versions of a specific package, making it impossible to satisfy all requirements. In other words, it is not possible to fulfill contradictory demands for dependent packages, making the installation of the packages they depend on impossible. In practice, the term "dependency hell" is often used to describe these problems [9]. Therefore, it is crucial for distributions/package managers to support parallel installation of multiple package versions.

A significant feature of certain package managers is the capability of sandboxing, i.e., isolating packages in separate environments [10]. This means that when such packages are executed, they do not have default access to the rest of the file system, hardware, and network resources. This enhances the security of the system, especially when software packages come from untrusted sources. However, explicitly granting permissions can be challenging in IoT environments. Additionally, packages often undergo rigorous testing and control in development/testing environments, making this feature less significant.

The unit of software that these managers work with is software packages, typically a single archive file that contains all the files belonging to the installed software. When updating a package version, all files of the currently installed version are removed by default, and then the files of the new version are unpacked. This approach offers a relatively high level of automation and reliability but lacks precision. Due to the weak granularity, a significant portion of package content is unnecessarily removed and reinstalled. This means that storage, communication, CPU, and memory resources are wasted, and system availability is reduced for the sake of replacing files with identical copies during version upgrades. This poses a minor problem in data centers, but it can have significantly more negative consequences in IoT environments.

2.4 Containers

Containers are another way of distributing and deploying applications that has gained immense popularity in the last decade. Containerized applications run in an isolated, virtual environment [11]. Unlike other types of virtualization, containers consume fewer resources and offer better performance. Furthermore, the isolation of containers solves the problem of version conflicts among different applications that use the same packages. This approach also enables achieving a high level of security since, by default, the application does not have access to the underlying file system and other resources. The main drawback of this approach usually lies in the size of the container, as they often contain duplicates of dependent software, in relation to either other containers or the host system.

3 Specificities of (A)IoT Environments

General software distribution systems are designed to, in principle, meet the general needs of various types of computer systems, such as personal computers, laptops and other portable devices, servers in data centers and cloud environments, etc. However, (A)IoT systems have specific characteristics related to hardware/software/network configurations, diverse software usage, communication with the environment, etc., which require specific considerations related to the distribution of software used on them.

3.1 Limited Internet and Network Connectivity

(A)IoT systems often have very limited Internet connectivity in terms of data transmission speed, allowed data quantity, stability, and reliability of the connection, etc. Additionally, it is not uncommon for the Internet connection to be unavailable altogether, and the network interfaces of the devices themselves have very limited capabilities (e.g., due to the intended minimal functional communication needs or to improve energy efficiency). In some configurations, each device has an independent Internet connection (e.g., using 3G/4G/5G networks), while in others, the entire (A)IoT device network accesses the Internet through a shared gateway. These devices are frequently used in remote rural areas where the only option for Internet access is through wireless technologies with limited speed and reliability. All of this makes downloading large software packages and their updates via the Internet very impractical, and often impossible in these environments.

3.2 Limited Processing Resources

Considering energy efficiency, thermal efficiency, and the target low cost of devices as some of the key criteria in the development of (A)IoT devices, it is understandable why their processing capabilities are scaled according to the projected needs. In other words, these devices usually do not have a significant amount of spare processing resources that could be utilized for installing downloaded software—decompressing archives, checking file integrity, replacing files in the file system, etc.

3.3 Limited Memory/Storage Resources

Installing or updating of large software packages requires significant memory (internal and external) resources. By default, a certain capacity of external memory is necessary to store the content of that package. However, the process of installation/update of package files itself demands additional storage capacity, first for downloading the archive files of the package, then for their decompression, and finally for copying them to the destination location. A crucial specificity of (A)IoT devices is that they predominantly use eMMC or SSD memory, which, besides general limitations in terms of capacity and read/write speed, also has a restriction on the number of possible value changes of memory cells.

Device manufacturers often provide expected memory endurance in years in their specifications, but such estimates are strictly related to the reference model of usage. In other words, the lifespan of the memory can be significantly longer or shorter depending on whether the content of the external memory is changed less or more intensively. Another complicating factor in this domain for (A)IoT devices is that their external memory is often soldered to the motherboard, making the replacement of worn-out devices for external memory considerably difficult or even impossible.

Therefore, in this regard, software distribution systems on (A)IoT systems should provide maximum efficiency in terms of minimizing the consumption of cycles of external memory cells, i.e., they should avoid any unnecessary file writing to the devices themselves.

3.4 Limited Set of Software Used

(A)IoT systems generally have well-defined functions in the environments they are installed in, which they execute repetitively. This means that they usually do not require a wide variety of installed software. Even in situations where they need large software packages, they typically utilize only a small subset of their functions and other resources. This is particularly evident in AIoT devices, where artificial intelligence functions are usually found within extensive software libraries. In this regard, software distribution systems should offer greater granularity during installation, allowing the possibility of omitting resources that will not be used.

However, this is a challenging task, as it is often impossible to predict precisely which parts of the software package will be used. Omitting components that will eventually be utilized may lead to the system ceasing to function properly.

4 Analysis and Results

This section of the paper presents the details and results of the analysis we conducted, aimed at determining the efficiency of a typical software distribution system—the package manager of a Linux distribution. The analysis considered the sizes and numbers of installed packages, as well as the new versions used for system updates. The number of files in the packages, their sizes, and the percentage of files and content that changed between versions were examined, specifically the proportion of modified content in the total size of distributed software. Additionally, we analyzed the changes between two versions of the Linux kernel source code that were covered by the official updates in this distribution.

4.1 Slackware Linux Distribution

The Slackware distribution is known for its proverbial stability, security, reliability, and simplicity of tools for its administration [12]. It is a distribution with 30 years of tradition, making it the oldest Linux distribution still in use. On the other hand, since it requires a high level of knowledge of Linux/UNIX operating systems for efficient use, it is typically not the choice for new users.

There are several reasons why we chose this distribution for demonstrating the issues addressed in this chapter. First, this distribution is an excellent choice for (A)IoT devices as it provides maximum system configuration capabilities, allowing optimal utilization of its resources. Second, its package management system is very simple and straightforward, enabling precise measurement of consumed resources and other parameters relevant to this research. Lastly, its 30-year lifespan signifies an exceptionally long and uninterrupted series of sound strategic decisions made by its creators, primarily Mr. Patrick Volkerding. Of course, considering the strict adherence to relevant standards by the Slackware distribution, all the conclusions drawn will be generally applicable to other typical Linux distributions as well.

4.2 The Observed Period and Versions

We analyzed version 15.0 of the Slackware Linux system. The period considered for analysis was from the release of this version (February 2, 2022) until the moment when this analysis was initiated (March 29, 2023). The analysis was conducted for both the stable [13] and current [14] development branches. As for the Linux kernel, we compared vanilla source code versions 5.15.19 and 5.15.94, released on February 1 and February 14, 2023, respectively, with a gap of two weeks. Finally, the analysis also includes the web server running version 14.1 of Slackware Linux, which has been actively used since 2014.

4.3 Linux Package Management Statistics

Initially, version 15.0 of Slackware distribution includes 1594 packages (all packages from all groups are included, except for "extra" and "testing"). The total size of these (compressed) packages is 2988 MB, while their total size on external memory after installation (i.e., decompression) is 14,266 MB. The total number of files in all packages is 680,477.

During the observed period, in the "stable" development branch, 78 original packages were updated, and the packages were updated a total of 263 times. Within these updates, 79,044 files were replaced, and the total size of these files (in the latest version) is 3.5 GB. Out of these files, only 11% (8845) differ from the initial version, and their total size is only 21% (736 MB) of the total size of all files in the updated packages. In other words, with the classical approach in package management—replacing the entire package, including all its files, whether they are modified or not—70,199 files were unnecessarily removed and reinstalled (in identical form), an average of 3.37 times. In terms of file size, the total size (in the latest version) of unnecessarily removed and reinstalled files is about 2.7 GB, which means the content of 7.65 GB was unnecessarily erased and written again.

In the same period, in the "current" development branch, 1152 original packages were updated, and the packages were updated a total of 8182 times. Within these updates, 561,700 files were replaced, and the total size of these files (in the latest version) is 16.5 GB. Out of these files, only 18% (103,658) differ from the initial version, and their total size is only 21% (3.3 GB) of the total size of all files in the updated packages. In other words, with the classical approach in package management—replacing the entire package, including all its files, whether they are modified or not—458,042 files were unnecessarily removed and reinstalled (in identical form), an average of 7.1 times. In terms of file size, the total size (in the latest version) of unnecessarily removed and reinstalled files is about 13.2 GB, which means the content of 100 GB was unnecessarily erased and written again.

4.4 Linux Kernel Source Packages Differences

We compared the packages with the source code of the Linux kernel, versions 5.15.19 and 5.15.94. In both cases, the size of the packages was about 121 MB, and the size of the unpacked files was about 1.2 GB. Version 5.15.19 contains 80,679 files, while that number is 64 higher in the newer version (80,743). By analyzing the changes, we concluded that in the newer version, 7322 files were modified, which accounts for only 9%. Projected to the size of files on external memory, these results indicate that by updating the mentioned package, unnecessarily 1.08 GB of identical data was deleted and re-written.

5 Conclusion

In this chapter, we analyzed the efficiency of traditional software distribution systems, primarily package managers on the Linux operating system, from the perspective of specific constraints and usage patterns of (A)IoT systems. The results of the conducted analysis indicate that the traditional approach to software distribution can be highly inefficient, as often over 90% of files are unnecessarily removed from the file system and then re-written with identical content. Such inefficiency is less significant in computer systems that have abundant local resources, fast and reliable Internet connections, and do not need to consider energy efficiency and external memory cycle consumption. On the other hand, all the mentioned factors commonly represent prominent constraints in (A)IoT systems, making the inefficiency of the traditional software update approach very significant in their case. The chapter does not provide a proposed solution, nor does it include an analysis of the utilization of distributed or installed software, which is a subject for future work.

Acknowledgments This article is based upon work from COST actions 19121 Network on Privacy-Aware Audio- and Video-Based Applications for Active and Assisted Living, 17124 Digital forensics: evidence analysis via intelligent systems and practices, supported by COST (European Cooperation in Science and Technology), www.cost.eu.

References

1. A. Katsman, Linux and the Internet of Things. Linux J. **2015**(255), 2 (2015)
2. S. Zhu, K. Ota, M. Dong, Energy-efficient artificial intelligence of things with intelligent edge. IEEE Internet Things J. **9**(10), 7525–7532 (2022)
3. A. Yazdanbakhsh, B. Akin, K.K. Seshadri, An evaluation of edge TPU accelerators for convolutional neural networks (2021)
4. F. Mancinelli, J. Boender, R. Cosmo, J. Vouillon, B. Durak, X. Leroy, R. Treinen, Managing the complexity of large free and open source package-based software distributions, in *21st IEEE/ACM International Conference on Automated Software Engineering (ASE'06)* (2006), pp. 199–208
5. F. costa Padilla, E. Baccelli, T. Eichinger, K. Schleiser, The future of IoT software must be updated, in *IAB Workshop on Internet of Things Software Update (IoTSU)* (2016)
6. Y. Zhang, V. Paxson, Detecting backdoors, in *9th USENIX Security Symposium (USENIX Security 00)* (2000)
7. https://flatpak.org/
8. https://snapcraft.io/
9. G. Fan, C. Wang, R. Wu, X. Xiao, Q. Shi, C. Zhang, Escaping dependency hell: finding build dependency errors with the unified dependency graph, in *Proceedings of the 29th ACM SIGSOFT International Symposium on Software Testing and Analysis* (2020), pp. 463–474
10. M. Páleník, Qubes OS
11. C. Pahl, Containerization and the PaaS Cloud. IEEE Cloud Comput. **2**(3), 24–31 (2015)
12. http://www.slackware.com
13. http://www.slackware.com/changelog/stable.php?cpu=x86_64
14. http://www.slackware.com/changelog/current.php?cpu=x86_64

Survey of Cybersecurity Risks in Online Gambling Industry

Mario Musa, Ivan Cvitić, and Dragan Peraković

1 Introduction

Complex systems represent a scientific field that studies how the elements of a system contribute to the collective behavior of the system and how the system interacts with its environment. Qualitatively, in order to understand the behavior of a complex system, it is necessary to comprehend not only the behavior of its individual components but also how they work together to shape the behavior of the whole. Complex systems and their desired behavior typically involve adaptability, self-decision making, robustness, decentralization, flexibility, and speed. Recently, there has been a greater focus on the structural characteristics of complex systems in the literature. These systems are increasingly described as networks characterized by decentralization, non-hierarchical, amorphous, dispersed, and distributed features. Complex systems, as networks of interconnected entities, are empirically studied through the rapidly growing volume of data available in various domains [1].

Analogously to the aforementioned ideas, today's information and communication systems are becoming highly complex and dynamic. Such systems involve a large number of heterogeneous and interconnected resources (at both logical and physical levels) that may be exposed to multiple security vulnerabilities and weaknesses. The dynamism of such systems is particularly prominent in modern information and communication systems, as new devices, means, and configurations

M. Musa
Hrvatska lutrija d.o.o, Zagreb, Croatia
e-mail: mario.musa@lutrija.hr

I. Cvitić · D. Peraković (✉)
University of Zagreb, Faculty of Traffic and Transport Sciences, Zagreb, Croatia
e-mail: ivan.cvitic@fpz.unizg.hr; dragan.perakovic@fpz.unizg.hr

L. Knapčíková, D. Peraković (eds.), *8th EAI International Conference on Management of Manufacturing Systems*, EAI/Springer Innovations in Communication and Computing, https://doi.org/10.1007/978-3-031-53161-3_8

can be constantly added, updated, and removed from the system, leading to new security vulnerabilities that did not exist at the time of design [2].

Examples of complexity and dynamism are evident in information and communication systems in domains such as critical infrastructure, finance, industry, and healthcare.

The aim of this paper is to identify and analyze cyber risks and security challenges in the environment of the online gambling industry, with Republic of Croatia "Croatian Lottery" case study as the biggest gambling company, which correspond to the definition of complex and dynamic information and communication systems within the concept of Industry 4.0. In the process of identification of cyber risks and security challenges, only publicly available data were used.

2 Analysis of Current State of Gambling Industry in Republic of Croatia

The exclusive right to organize games of chance in the Republic of Croatia belongs to Croatian lottery Ltd. Other companies based in the Republic of Croatia may also be allowed to organize casino games and betting games by the decision of the Government and approval from the Ministry of Finance, following a public tender process. In 2007, the first concessions for organizing games of chance in automated clubs were approved, and by the end of 2014, the Ministry of Finance and the Government allowed for 29 new concessions (increasing from 19 to 48). After the financial crisis in 2010, the gambling market experienced growth and recovery. In addition to regulatory changes in 2014, the development of digitalization has greatly influenced market growth, particularly in the online gambling segment [3]. According to statistical data, at the end of 2016, there were 2500 betting shops, 15 casinos in 12 cities, 219 automated clubs, "State lottery," and at least five online casinos operating in the market. In 2015, the first legal online casino was launched [4] (Fig. 1).

According to Fig. 2, there is a visible growth in the annual revenue from online gambling games in Europe from 2003 to 2023. It is predicted that the revenue in the online segment will reach 33 billion euros in 2023.

According to Statista, the share of online gambling games in the total number of gambling games was significant in many EU countries in 2019. In Croatia, it accounted for 23.4%, which is higher than countries such as France (19.2%), Germany (19.2%), Spain (12.2%), or Italy (10.6%).

The prevalence trend of online gambling games in Europe is consistently upward, which is also reflected in the trend in the Republic of Croatia, with predictions for the future period supported by the observed increase thus far. Corresponding to the demonstrated growth in popularity and the utilization of the Internet as a platform for gambling, there is a causal emergence of an increasing number of cyber threats targeting the systems and end users of online gambling games. The issue of cyber threats and the risks associated with successful implementation of various

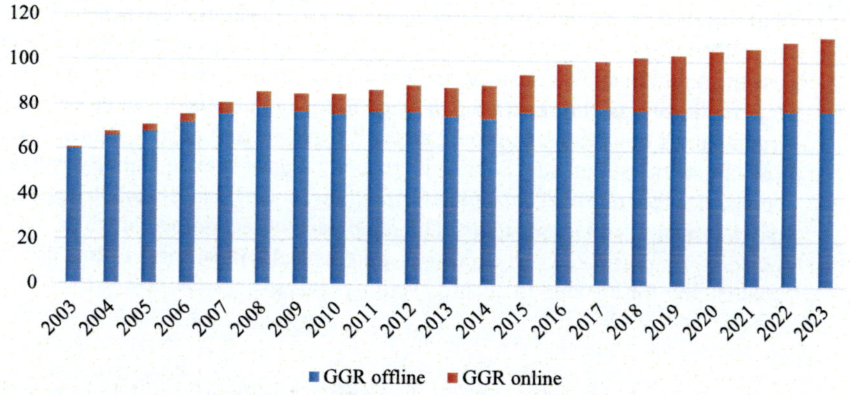

Fig. 1 Share of online gambling games in the total revenue from such games

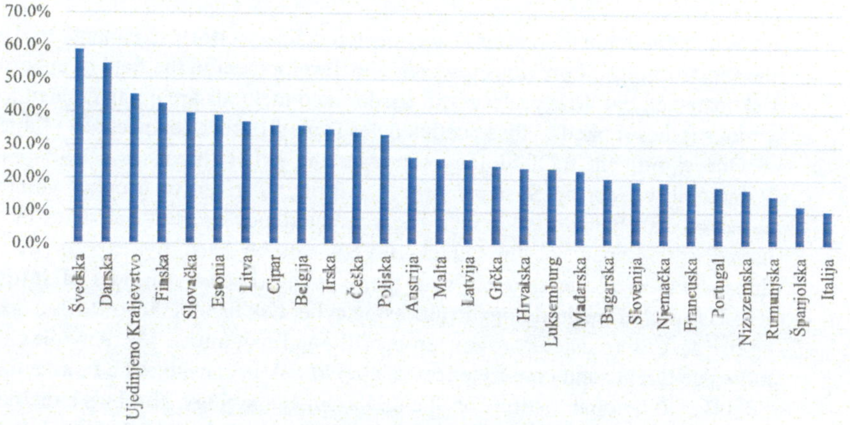

Fig. 2 The share of online gambling games in the total number of gambling games

attacks are becoming increasingly prominent in this industry segment, necessitating a systematic approach to addressing and resolving such problems. To this end, it is essential to identify cyber threats targeting online gambling games and provide suggestions and guidelines for their timely and adequate mitigation.

3 Identification of Security Challenges in the Environment of Online Gambling

Cyber risk can be defined as the financial loss or reputational damage resulting from failures in an organization's IT systems. Global cybercrime cost approximately $600 billion in 2017. According to IBM's security report in 2020, unauthorized data disclosure represents an average loss of $3.92 million per incident for a company. Cyber risks are dynamic due to ongoing digital innovations, increased use of Internet-enabled devices, and the current sophistication of hackers. Business interruption costs and reputation damage following an attack can be detrimental to a company's operations. Financial services firms are particularly vulnerable and are key targets of cybercrime. Sensitivity to cyberattacks is exponentially increasing as companies rely more on data, cloud computing, social media, mobile devices, third-party software, and outsourcing. The rapid pace of technological advancement, the potential for correlated risk exposures, and the lack of historical claims data make it challenging for insurers to quantify future losses [5].

3.1 Overview of Currently Available Methods for Identification and Classification of Cybersecurity Risk

Currently, there are numerous academic and industry efforts to enhance and address the shortcomings of previous research and development in the field of cybersecurity risk management. Many successful models and methods are being created to tackle various issues related to these deficiencies from different perspectives (Table 1).

One significant methodology for assessing cyber threats is CVSS (Common Vulnerability Scoring System), where a rating is given to a cyber vulnerability based on its characteristics [6]. The characteristics used to calculate the scores include the impact of the attack. CVSS can be used by experts to identify threats/vulnerabilities and evaluate them. FAIR (Factor Analysis of Information Risk) is another methodology in assessing cyber risk factors that utilizes a taxonomy of risk factors to calculate their probability and magnitude [6]. Anything that can influence threats and consequences related to risk is considered a risk factor. NIST SP800-39 (National Institute of Standards and Technology, NIST) is a methodology for assessing risks in government information and communication systems and organizations [7]. OCTAVE (Operationally Critical Threat, Asset, and Vulnerability Evaluation) is a framework for identifying risky assets, threats, and vulnerabilities that can exploit those threats [8]. OCTAVE is used for both medium and small enterprises. This method has three phases of a progressive work framework. Organizational and technological risk profiles are reviewed in the first and second phases of OCTAVE. Mitigation strategies and plans are developed in the final phase. Another risk assessment and mitigation framework called TARA (Threat Agent Risk Assessment), designed by Intel, assesses cyber threats and selects the most important measures for their mitigation [9]. CORAS is another method of security risk analysis that attempts to identify threats and vulnerabilities using a modeling-based approach [10]. This approach benefits from modeling tools based on language and UML. OWASP (Open Web Application Security Project) provides standards, security checklists, and procedures for web-based systems [11].

Table 1 Review of cyber risk assessment frameworks and methodologies [6]

Framework/methodology	Brief description	Advantages	Disadvantages
CVSS (*Common Vulnerability Scoring System*)	Standard for assessing vulnerability severity on a scale of 1–10	Comprehensive vulnerability verification system	Not designed for risk classification
FAIR (*Factor Analysis of Information Risk*)	A methodology for quantifying risk-related factors	Provides a taxonomy of risk factors with a methodology for their quantification	Not a risk assessment method
NIS SP800	NIST is framework for cybersecurity risks. NIST SP800 provides comprehensive countermeasures/checklists for cyber threats	Can be used by companies to define appropriate controls related to cyber vulnerabilities	Does not provide a way to quantify the effectiveness of control
OCTAVE (*Operationally Critical Threat, Asset and Vulnerability Evaluation*)	A comprehensive procedure for risk assessment and evaluation through four phases, from selecting risk measurement criteria to identifying approaches for mitigation	All processes of assessing cyber risks and evaluation are covered by this framework	Does not provide detailed guidance for each step of each phase, which may be problematic in different industries
CMMI (*Capability Maturity Model Integration*)	A maturity model with five levels that guides organizations through best practices for continuous improvement of production methods, especially in software development	A comprehensive process that oversees all aspects of production maturity	Risk management is considered a step in the third level of maturity, but specific obligations are not considered for the selected method
TARA (*Threat Agent Risk Assessment*)	A methodology for identifying, ranking, and filtering threats, prioritizing mitigations based on threat importance	Well-structured with knowledge libraries for threat and exposure identification	Not designed for risk classification
CORAS	A framework for security risk analysis with programming language and tools based on the Unified Modeling Language (UML)	Provides step-by-step processes from threat and risk identification to response	Does not go into detail on how to find threats. The utilities and documents of this method have not been updated since 2014
IoTMM (*Internet of Things Maturity Model*)	A maturity model for the Internet of Things with five levels	Focuses on IoT design and implementation	(1) Lacks a precise risk assessment methodology. (2) Reliable service quality is mentioned in the third level of maturity
OWASP (*Open Web Application Security Project*)	Documents, checklists, and recommendations in the field of web application security	Comprehensive overview of all production threats/vulnerabilities	Focuses on web and threats rather than risks/events

3.2 *Identification and Classification of Cyber Threats in the Online Gambling Industry*

The increased usage of digital services is directly linked to the opportunities for cybercrime. The online gambling market is an attractive target for numerous cyber threats associated with financial theft or personal data breaches. It is important for gambling operators to prioritize cyber risk and establish a secure platform for their users. GDPR has played a significant role in ensuring consumer data privacy and must be taken seriously and implemented by online gambling service providers.

Similar to financial risk management, cyber risk management plays a crucial role in ensuring business resilience. These risks encompass internal threats, web application security challenges, and issues related to ransomware [12].

The cyber threats and risks addressed in this study are the result of applying diverse OSINT (Open Source Intelligence) methods and tools at the network infrastructure level, as well as web-based applications and environments. Additionally, the research covers cyber vulnerabilities and weaknesses, including threats related to user identification and authentication in online gambling systems, as well as threats targeting users and employees of such systems.

Level of Network Infrastructure and Public Exposure of Sensitive Information The online gambling system of the Croatian Lottery is focused on providing the opportunity to play numerous games through web browsers using various web applications. It is essential to be aware of and monitor the suitability of public exposure of individual servers within the Croatian Lottery system.

Server name	IP address	Type	Reverse DNS	Block owner
hlz1appliance1.lutrija.hr	195.29.226.5	A	hlz1appliance1.lutrija.hr	T-HT Croatian Telecom Inc.
z1appliancehl1.lutrija.hr	195.29.226.5	A	z1appliancehl1.lutrija.hr	T-HT Croatian Telecom Inc.
lutrijadns1.lutrija.hr	195.29.226.10	A	lutrijadns1.lutrija.hr	T-HT Croatian Telecom Inc.
hlz1appliance2.lutrija.hr	195.29.226.12	A	z1appliancehl2.lutrija.hr	T-HT Croatian Telecom Inc.
z1appliancehl2.lutrija.hr	195.29.226.12	A	hlz1appliance2.lutrija.hr	T-HT Croatian Telecom Inc.
lutrijadns2.lutrija.hr	195.29.147.25	A	lutrijadns2.lutrija.hr	T-HT Croatian Telecom Inc.

Table 2 presents an example list of publicly accessible servers with their IP addresses and accompanying information that can be useful when planning unauthorized activities by malicious users.

Table 2 Summary results of security assessment

		Trustworthiness				
		User confirmed	High	Medium	Low	Total
Risk	High	0	0	0	0	0
		0.00%	0.00%	0.00%	0.00%	0.00%
	Medium	0	0	1779	0	1779
		0.00%	0.00%	36.10%	0.00%	36.10%
	Low	0	0	2479	19	2698
		0.00%	0.00%	54.30%	0.40%	54.70%
	Informative	0	0	0	457	457
		0.00%	0.00%	0.00%	9.30%	9.30%
	Total	0	0	4458	476	4934
		0.00%	0.00%	90.40%	9.60%	100.00%

Level of Internet Domain and Web Applications

Due to the primary channel of providing gambling services through web-based applications, potential cybersecurity risks need to be analyzed within the framework of the OWASP Top 10 document. The goal of the OWASP organization is to enable the development and maintenance of applications while maintaining the required level of security. The OWASP Top 10 project was initiated to increase awareness of application security by identifying key risks faced by organizations. The project is referenced by many standards, literature, tools, and organizations. It was launched in 2003, and the latest version was published in 2021. OWASP Top 10 focuses on identifying the most critical risks for a wide range of organizations. Generic information about the likelihood and impact of each risk is provided based on a risk assessment methodology derived from OWASP.

In order to check for potential cybersecurity risks within the web applications and domain of the information system of Croatian Lottery, a security assessment and vulnerability analysis were conducted using the OWASP ZAP 2.11.0 tool. For the purpose of this research, the Internet domain http://www.lutrija.hr, including its subdomains and all associated hyperlinks, were scanned as they represent publicly exposed resources accessible to all users (legitimate and illegitimate).

The summary results of the security assessment are presented in Table 2, which shows the level of risks identified during the security assessment. The methodology used for assessing cyber risks aligns with the OWASP risk assessment methodology [13]. From the presented table, it is evident that the primary and associated Internet domains did not exhibit high risks in terms of vulnerabilities. The security assessment identified a higher number of risks classified as medium, low, and informational. Although they are not classified as high risks, these identified risks need to be adequately addressed and minimized. The risks identified during the security assessment align with the top 10 risks for web applications according to the OWASP methodology. An example of a medium security risk identified during the security assessment is Cross-Domain Misconfiguration, as depicted in Fig. 3. The

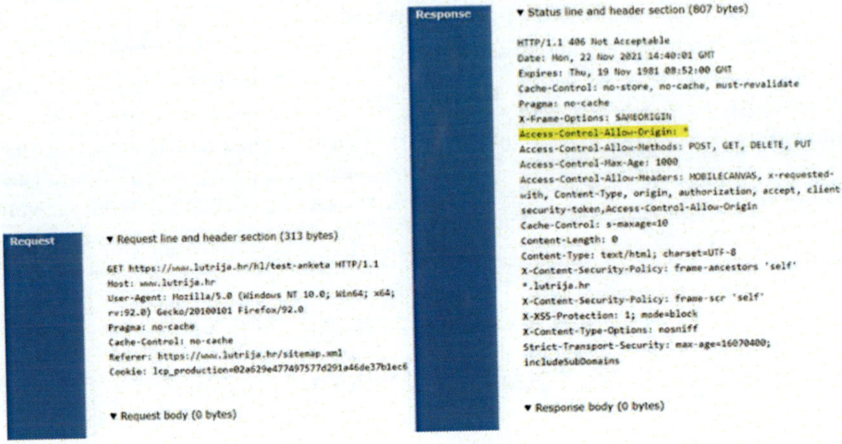

Fig. 3 An example of an identified cyber vulnerability under the risk category of Broken Access Control

identified risk ranks first on the OWASP Top 10 risk list for 2021 under the Broken Access Control risk category.

Access control enforces rules that prevent users from acting outside their designated permissions. Failed implementation of access control typically leads to unauthorized disclosure of information, alteration or destruction of data, or performing business functions beyond user limitations.

Common access control vulnerabilities include the following:

- Violation of the principle of least privilege or denying by default, where access should only be granted to specific capabilities, roles, or users, but is available to everyone.
- Bypassing access control checks by modifying URLs (unauthorized parameter tampering or forced browsing), application state, or HTML pages or by using attack tools that modify API requests.
- Allowing viewing or editing of other accounts by providing a unique identifier (insecure direct object references).
- Accessing APIs without proper access controls for POST, PUT, and DELETE.
- Privilege escalation, acting as an unauthenticated user or acting as an administrator when logged in as a regular user.
- Manipulating metadata, such as replaying, unauthorized modification of access control tokens like JSON Web Tokens (JWT), cookies, or hidden fields for privilege elevation or JWT abuse.
- Misconfiguration of Cross-Origin Resource Sharing (CORS), enabling API access from unauthorized/untrusted sources.
- Forced browsing of pages that require authorization by unauthorized users.

The Level of Cyber Risks Targeted at End Users of Online Gambling Systems The user, as a component of the information and communication system, often represents the weakest point in terms of cyber risks. In the online gaming environment, numerous cyber risks can be observed at the user level, which frequently occur and have a significant negative impact on users, including privacy breaches, theft of personal data or money. Some common cyber threats in the online gambling environment include phishing attacks, distributed denial-of-service (DDoS) attacks, and man-in-the-middle (MitM) attacks [14].

Phishing attacks involve unauthorized users attempting to gather user credentials for system access by exploiting the psychological factor (naivety) of users, luring them to a fake web destination for system access.

DDoS attacks aim to deny legitimate users access to the system by generating a large number of illegitimate requests, thus exhausting the system's available resources for serving requests.

MitM attacks involve unauthorized users intercepting and modifying the communication between a legitimate user and the system they are accessing. This type of threat opens up the possibility of various additional attacks, such as stealing digital certificates for system access, altering DNS requests and redirecting users to other web destinations, or enticing users to download ransomware or visit mirrored websites aimed at stealing their credentials.

Although the set of cyber threats consists of a large number of possible attacks, here we have identified and listed those threats that, given the observed environment and the type of information and communication service provided to users by the online gambling system, represent significant security risks.

4 Guidelines for Prevention and Mitigation of Identified Cyber Threats

4.1 Guidelines for Improvements at the Network Infrastructure Level

At the network infrastructure level, infrastructure resources that are visible through various levels of network scanning and exposed to the public communication network are presented. Information about services (server names, types of network services, network applications, and open communication ports) can pose a security risk in terms of mapping the communication network of the system and identifying the starting point for various cyberattacks, regardless of whether the resources are located in the DMZ zone. It is strongly suggested that an organization always maintains a clear picture of publicly available system information (digital footprint) and reduces the exposure of unnecessary network resources to the public communication network as needed.

To enhance clarity and comprehensiveness in the realm of publicly accessible information, it is imperative to conduct further, in-depth research using the Open Source Intelligence (OSINT) framework. This approach necessitates the adoption of its ongoing application to ensure a thorough understanding of the available data scope. These activities achieve the following benefits:

- Timely alerts about the leakage of sensitive data
- Timely insight into oversharing of sensitive data through social media
- Detection of vulnerabilities in systems exposed to the public communication network that could lead to a security incident
- Discovery of different versions of documents that exist on the Internet, which have been mistakenly published and contain sensitive information

The primary guideline for reducing the digital or OSINT footprint is proactive action and ensuring that the organization exposes as little sensitive information as possible. It is necessary to provide adequate education for employees at all levels in the field of cybersecurity, with an emphasis on reducing the exposure of their own and organizational sensitive data.

4.2 Guidelines for Improving the Security of the Internet Domain Level and Web Applications

In accordance with the security validation conducted using the OWASP methodology, certain cyber risks have been identified at the level of the Internet domain www.lutrija.hr, its associated subdomains, and corresponding hyperlinks. For the presented examples of identified cyber risks, it is suggested to take action and implement appropriate solutions to minimize vulnerabilities.

For the identified example of a moderate security risk related to cross-domain misconfiguration, as shown in Fig. 3, it is necessary to ensure that sensitive data is not accessed in an unauthorized manner, for example, by implementing IP whitelisting. The HTTP header Access-Control-Allow-Origin should be configured to a more restrictive set of domains or completely remove all CORS headers to enforce a more restrictive Same Origin Policy (SOP) implementation by the web browser.

An example attack can be observed through the following scenario. An unauthorized user attempts to access an area that requires specific authorization (e.g., https://example.com/app/getappInfo) or administrative privileges (e.g., https://example.com/app/admin_getappInfo) by manipulating the URL. The identified vulnerability exists if an unauthorized user successfully accesses the URL: https://example.com/app/getappInfo or an authorized user without administrative privileges successfully accesses the URL: https://example.com/app/admin_getappInfo

Access control is only effective in reliable server-side code or serverless API where the attacker cannot modify access control checks or metadata. To prevent the identified vulnerabilities, the following measures are recommended:

The access control mechanisms should be implemented consistently throughout the application, including minimizing the use of shared CORS services. The access control model should enforce ownership over records rather than allowing users to create, read, update, or delete records. Disable directory listing on the web server and ensure that metadata and backup data are not exposed in root server directories. Log attempts to violate access control rules and promptly notify administrators. Limit the rate of access to APIs and controllers to mitigate the damage from automated attack tools.

Stateful session identifiers should be invalidated on the server after session termination. Stateless JSON Web Tokens (JWT) should have short lifetimes to minimize the space for malicious activities. For longer-lived JWT tokens, it is recommended to follow OAuth standards for access token revocation.

To prevent the identified vulnerabilities, secure installation processes should be implemented, including the following:

- A repeatable hardening process that enables quick and easy deployment of a properly secured alternative environment. Development, QA, and production environments should be identically configured with different credentials used in each environment. This process should be automated to minimize the effort required for setting up a new secure environment.
- A minimal platform without any unnecessary features, components, documentation, or samples. Unused features and frameworks should be removed or not installed.
- Task reviews and updates of configurations relevant to all security advisories, updates, and patches as part of the management process. Consider permissions for cloud storage (e.g., S3 bucket permissions).
- A segmented application architecture provides effective and secure separation between components or users, with segmentation, containerization, or cloud security groups (access control lists).

In addition to the aforementioned recommendations for improving the security of the web environment, it is recommended to perform additional security testing of web applications and web services using the OWASP WSTG (The Web Security Testing Guide) framework. By applying this framework, not only the technical implementation of specific web-based solutions is evaluated but also the adequacy of employee education and awareness of cyber threats, as well as the examination of policies and processes and their suitability and clarity of enforcement.

Considering the sensitivity of online gambling in terms of transactions and the handling of sensitive data, it is further proposed to implement the ASVS (Application Verification Standard) level 2 standard in future security enhancement phases. Level 2 ensures that security controls are established, effective, and utilized within the application. It is suitable for applications that handle significant B2B transactions, including those processing medical and healthcare information or other sensitive assets, or industries where integrity is a critical aspect of business protection, such as the gaming industry, in order to prevent fraud in online games.

4.3 Guidelines for Improving the Security of Online Gambling Users

In order to minimize cyber risks identified at the level of online gambling users, the implementation of multiple interconnected measures is proposed. The first measure, aiming to raise users' awareness of cyber threats (referred to as cybersecurity awareness), involves continuous awareness-raising through targeted and focused education. This measure entails the design and development of easily understandable documents (such as infographics or similar materials) on the most common cyber threats users may encounter when using online gambling systems, how to recognize them, and how to adequately protect themselves. Distribution and dissemination of this type of education can be carried out through existing communication channels such as organizational websites, digital newsletters, and the like. The implementation of this measure would positively impact the prevention of cyber threats such as phishing, Man-in-the-Middle (MitM) attacks, unauthorized redirection to unverified and malicious Internet destinations, as well as DNS session hijacking.

Furthermore, the implementation and enforcement of multifactor authentication are recommended for users. Multifactor authentication utilizes at least two authentication methods, typically based on two conditions/attributes: (1) something the user knows (username and password) and (2) something the user possesses (mobile device). Optionally, a third attribute can be used, something the user is (biometric characteristics such as fingerprint), which is not necessary at this level of user authentication and authorization. The implementation of this method would further reduce the risk level of the aforementioned cyber threats (phishing and MitM).

These mentioned methods allow for the mitigation of cyber risks by reducing the likelihood of phishing and MitM cyberattacks. However, they do not affect the likelihood of cyber threats targeting the availability of the system, such as various Distributed Denial-of-Service (DDoS) attacks. To ensure continuous service availability to end users, methods and solutions for protecting and mitigating this type of attack, which is continuously increasing in trend due to its simplicity and effectiveness, need to be applied. Due to the nature of the service and the number of users, online gambling systems face an increased risk of DDoS attacks. The implementation of existing solutions such as Cloudflare Security, Imperva Incapsula, or F5 Silverline DDoS Protection is proposed. According to numerous sources and research, the cost of system unavailability caused by DDoS attacks ranges from $5500 per minute for Small and Medium Enterprises (SMEs), while such attacks can be rented as a service (DDoSaaS) for a price of $10 per hour [15]. The risk has been further amplified by the pandemic, which has witnessed an increase in online services like gambling, resulting in a 350–550% rise in the number of DDoS attacks compared to the pre-pandemic period [16]. The implementation of existing solutions also opens up opportunities for research and development of new detection and response models for DDoS attacks in the online gambling environment.

5 Conclusion

In this study, a comprehensive analysis of cyber risks in the information and communication system of online gambling was conducted, covering various levels of complexity. The analysis focused on network infrastructure, Internet domain, web applications, and end users, identifying cyber vulnerabilities and associated threats. The use of tools and the OSINT methodology provided valuable insights into publicly available data, highlighting the need for organizations to have a systematic understanding of their public exposure and implement continuous OSINT analysis to minimize risks.

At the Internet domain and web application levels, security testing based on the OWASP methodology revealed medium-level risks, particularly in cross-domain misconfiguration and content security policies. To mitigate these risks, implementing ASVS level 2 and conducting regular security testing using the OWASP WSTG framework are recommended.

Additionally, this study emphasized the importance of addressing cyber threats that can compromise user privacy, sensitive data, and service availability in online gambling. Measures such as developing educational materials to raise awareness among users about cyber threats, implementing multifactor authentication to protect user accounts, and deploying DDoS protection solutions to ensure service availability were proposed.

By implementing these measures, organizations can enhance the security posture of online gambling systems, minimize cyber risks, and safeguard the privacy and experiences of users. However, it is crucial to maintain vigilance, conduct ongoing risk assessments, and adapt security measures as new threats emerge in the ever-evolving cyber landscape.

References

1. P. Antoniou, A. Pitsillides, *Understanding Complex Systems: A Communication Networks Perspective* (Department of Computer Science, University of Cyprus, 2007), pp. 1–22
2. N. Kheir, A.R. Mahjoub, M.Y. Naghmouchi, N. Perrot, J.P. Wary, Assessing the risk of complex ICT systems. Ann. Telecommun./Ann. Telecommun. **73**, 95–109 (2018). https://doi.org/10.1007/s12243-017-0617-0
3. H. Šimović, A. Bajo, M. Davidović, F. Jelavić, Gambling market in Croatia: Financial performance and fiscal effect. Fiscus (2019). https://doi.org/10.3326/efiscus.2019.9

4. A. Raspor, I. Bulatović, A. Stranjančević, D. Lacmanović, How important is gambling in national GDP: Case study from Austria, Croatia, Italy and Slovenia. Economics **7**, 31–49 (2019). https://doi.org/10.2478/eoik-2019-0004
5. H. Bahşi, U. Franke, E.L. Friberg, The cyber-insurance market in Norway. Inf. Comput. Secur. **28**, 54–67 (2019). https://doi.org/10.1108/ICS-01-2019-0012
6. B. Sheehan, F. Murphy, A.N. Kia, R. Kiely, A quantitative bow-tie cyber risk classification and assessment framework. J. Risk Res. **0**, 1–20 (2021). https://doi.org/10.1080/13669877.2021.1900337
7. A. Nurliyani, D. Syamsuar, A. Haidar Mirza, Assessment IT risk management pada Laboratorium Teknik Komputer. J. Inf. Telecommun. Eng. **2**, 62–68 (2019)
8. C.J. Alberts, S.G. Behrens, R.D. Pethia, W.R. Wilson, *Operationally Critical Threat, Asset, and Vulnerability Framework, Evaluation* (1999). https://insights.sei.cmu.edu/documents/1210/1999_005_001_16769.pdf
9. J.E.Y. Rossebo, F. Fransen, E. Luiijf, Including threat actor capability and motivation in risk assessment for smart GRIDs, in *2016 Joint Workshop on Cyber-Physical Security and Resilience in Smart Grids (CPSR-SG)*, (IEEE, 2016), pp. 1–7. https://doi.org/10.1109/CPSRSG.2016.7684094
10. M.S. Lund, B. Solhaug, K. Stølen, *Model-Driven Risk Analysis* (Springer Berlin Heidelberg, Berlin, Heidelberg, 2011). https://doi.org/10.1007/978-3-642-12323-8
11. D. Kellezi, C. Boegelund, W. Meng, Towards secure open banking architecture: An evaluation with OWASP, in *Lecture Notes in Computer Science (Including Subseries Lecture Notes in Artificial Intelligence and Lecture Notes in Bioinformatics)*, (2019), pp. 185–198. https://doi.org/10.1007/978-3-030-36938-5_11
12. Cyphere, *Cyber Security in the Gambling Industry*, https://thecyphere.com/industries/gambling-cyber-security/. Last accessed 11 Oct 2021
13. OWASP, *OWASP Risk Rating Methodology*, https://owasp.org/www-community/OWASP_Risk_Rating_Methodology. Last accessed 11 Oct 2021
14. D. Peraković, L. Cetinić, I. Cvitić, M. Periša, *Cybersecurity Validation in the Online Gambling Industry*. Presented at the (2022). https://doi.org/10.1007/978-3-031-06025-0_7
15. M. Gomez, *Dark Web Price Index* (2020). https://www.privacyaffairs.com/dark-web-price-index-2020/. Last accessed 20 Oct 2021
16. Kaspersky, DDoS attacks against educational resources increased by more than 350% this spring, https://www.kaspersky.com/about/press-releases/2020_ddos-attacks-against-educational-resources-increased-by-more-than-350-this-spring. Last accessed 10 Oct 2020

Development of CIO Competencies

Mijo Šarčević

1 Introduction

The high-tech globalized world demands modern man to constantly adapt in all spheres of life and work. On the one hand, a globalized world offers great blessings, and on the other, huge challenges. The term *Fourth Industrial Revolution* is often mentioned in technological circles as the driver of economic growth and the development of society in general. It is based on the desire to achieve lower production costs and greater efficiency through a more significant integration of knowledge and resources, but how do we achieve this? Is technical knowledge and proficiency sufficient for those responsible in this process? How do we effectively transfer knowledge and understanding to those involved in the decision-making process? Understanding ICT (*Information and Communication Technologies*), their importance, and their application in concrete situations as well as in the context of creating added value is not an easy task. IT management is a very critical area in many organizations because the entire business relies more and more on ICT. Technologies such as BigData and IoT are sources of data that need to be used to create new values. To manage critical organizational resources, it is necessary to have a CIO (*Chief Information Officer*) who can communicate effectively [1] and clearly [2] within the organization as well as with partners and suppliers. This points to the fact that among all the skills a CIO must possess, his communication skills are the most important [3]. In the year 2023, which was declared by the European Commission as the European Year of Skills, the question of what skills are important for a successful CIO was raised. Is this, as we often think, predominantly their technical proficiency? This paper aims to raise awareness of the fact that

M. Šarčević (✉)
Kaufland Hrvatska k.d, Zagreb, Croatia

L. Knapčíková, D. Peraković (eds.), *8th EAI International Conference on Management of Manufacturing Systems*, EAI/Springer Innovations in Communication and Computing, https://doi.org/10.1007/978-3-031-53161-3_9

communication skills and their great importance for the success of projects in ICT fields are insufficiently valued.

2 Theoretical Framework

2.1 CIOs: Their Role in the Organization

IT manager and *CIO* are the most commonly used titles in the world for the person responsible for the ICT system of an organization. However, sources show many variations of this title, depending on the period, business, and geographical environment of its origin. The influence of earlier research by different authors is not irrelevant either. Table 1 shows a brief overview of the most common titles according to the authors who used them. Since there is no agreement on the title for such an important function, there is also no unified opinion about the most appropriate position of the CIO within the organizational structure [4, 5]. However, the CIO needs to be in the appropriate hierarchical position [6], as this gives him greater influence in the decision-making process.

It is not possible to explore the necessary skills of a CIO without defining his role in the organization. These skills changed in accordance with the development of society, technology, and the change of attitude about the importance of ICT for the success of an organization as well as society in general. They can be observed through several specific phases. In the early 1980s, Ives and Olson studied how the role of the information systems manager changed from a relatively unimportant service function to an important instrument for organizational changes and profit generation. The CIO is becoming a coordinator, motivator, and planner, thus emphasizing that he is more of a manager and less of a technologist [7] and that

Table 1 Overview of the CIO title

Different names for CIO	Authors
EDP – Electronic data processing manager	Ives & Olson, 1981
MIS – Management information systems manager	Ives & Olson, 1981 Merhout, Havelka, & Hick, 2009
I/S executive – Information systems executive	Rockart, Ball, & Bullen, 1982 Benjamin, Dickinson, & Rockart, 1985
CIO – Chief information officer	Synnott & Gruber, 1981 Stephens & Loughman1994 Weill & Ross, 2009 Waller Rubenstrunk, & Hallenbeck, 2010
CTO – Chief technology officer	Karanja & Zaveri, 2012
IT Director	Peppard & Ward, 1999 Boehm, Stolze, & Thomas, 2013
IS Manager – Information systems manager	Wu, Chen, & Chang, 2007 Mađarić, Krakar, & Lovrek, 2009

he is a senior executive manager responsible for establishing corporate information policies, standards, and managing all sources of information [8]. According to Chun and Mooney, in the 1980s, CIOs are behind-the-scenes technicians, while in the 1990s, they are becoming innovators who are trying to move up the organizational ladder. Simultaneously with the changes in strategy, the development of technology, and the needs of modern business processes, the role of the CIO also changed [9]. In the beginning, the CIOs were seen exclusively as a cost center, with a purely technical role and no influence on finances and management. In recent times, by accepting his role as an initiator and strategist [10] who creates new value, the CIO is becoming an important link in the changes that provide progress. While managing the IT sector, he can either be an executive manager dealing with strategy and processes or a technical manager in charge of efficient infrastructure [9]. It is not a new name for an old thing [11] because the modern CIO is becoming a management board member who manages matters of strategic importance for the organization he works for [12].

2.2 *The Importance of Communication Skills of CIOs*

The art of communication is a basic managerial skill [13], and it is especially important for people in higher hierarchical positions. Communication is the act or process of using words, sounds, signs, or behavior to convey and exchange information with someone or to express one's ideas, thoughts, and feelings [14]. The importance of communication between team members is best seen in the book of Genesis (Genesis, 11–11:9) where people try to build *a city and a tower with its top in the heavens*. Their work was going great while they were using the same language, but then Yahweh disrupted their work by confusing the language of the workers so that they could no longer understand one another. As a result, they stopped cooperating and building the tower [15]. The study of the communication skills of CIOs began in the 1970s, inspired by Mintzberg's conclusions that managers prefer verbal communication instead of formal reports [16]. Their role is very sensitive and demanding because they are constantly communicating with suppliers, colleagues, partners, and superiors, and this requires strong social skills. The absence of such skills can make it difficult for the CIO to communicate with colleagues as well as with other managers [17]. Mintzberg's theory was also confirmed by Ives and Olson who, through an analysis of the CIO's daily activities, concluded that oral communication makes up 76% of his working day [7]. One of the four critical success factors of a CIO is communication, and consequently, communication skills [18], because he needs to be a business-oriented manager with strong political, organizational, and communication skills [19]. In the period after the 1990s, the CIO is required to fulfill even more roles. In this context, the CIO must be *a communicator, teacher, motivator, leader, politician, relationship builder, and team player* [20]. When communicating, he must *use language carefully and avoid technical jargon as well as use the connotative value of words,* i.e., *he must possess strong communication skills; both verbal and non-verbal due to the increased number of contacts with the board and management* [4]. *He is* responsible for the success of the IT sector, and thus for the success of the organization in which he works. It is not enough to successfully achieve a goal, success must be communicated to the outside world [21] which consists of the management, ICT users, partners, and suppliers. At the beginning of the new millennium, Mintzberg's spokesperson role was among the most important roles [22], and CIOs were required to have an excellent understanding of business communications [23]. Mutual understanding between the CEO and the CIO is the most important aspect of their efforts to effectively exploit information technologies, but it also affects the financial contribution of ICT to the success of the organization [24]. A CIO should possess more skills than the executive director, for example, business knowledge, communication skills, and political wisdom [25]. He must possess a combination of ICT technical knowledge as well as business knowledge [26]. The constant development of ICT puts the communication role of the CIO in the foreground, or rather the need for effective communication [1]. And with the help of the aforementioned communication skills, the CIO provides the executive director with an understanding of ICT needs, which is a prerequisite for the effectiveness of his role in the organization [27]. The quality of his communication is crucial for the success of projects because if he does not communicate effectively, his projects will fail, either in the approval phase, when they are rejected by the management, or in the implementation phase when they are rejected by the users [28]. The arrival of the world economic crisis creates new requirements and demands adjustments from CIOs. The IT sector is required to constantly increase its efficiency, which is why the CIO must develop an effective management and communication strategy [29] – for he is the one who has to act as a link between the two parties – the management and ICT users, and the IT sector as the one whose success is measured by the satisfaction and efficiency of the former two. He does not have to be a technical guru as long as he spends most of his working day talking to other managers and determining and planning IT strategies to support the organization's objectives [30]. The 1990s were marked by numerous high-quality studies on communication skills and the role of CIOs in the organization, and the last decade confirmed how important it is to regularly discuss and examine the role of CIOs and their communication skills.

3 Research Methodology

The purpose of this research is to address the importance of the communication skills of CIOs in fulfilling their role in the organization. When choosing appropriate CIOs, it was necessary to fulfill several criteria according to the concept first developed by Stephens and Loughman, which are as follows:

1. He must be the highest-positioned person responsible for ICT in the organization.
2. Corresponds to a maximum of two hierarchical levels above himself.

3. Responsible for information systems and telecommunications, customer support.
4. Responsible for the strategic planning of ICT resources [4].

In the selection process, it was important that the organizations they work for are from different business areas and that the private and public sectors of the Republic of Croatia are evenly represented. Ten CIOs from various business areas such as higher education, production, construction, trade, insurance, and healthcare were interviewed.

3.1 *Qualitative Research Approach*

Using a qualitative method and an in-depth interview with the CIOs, an attempt was made to answer the question of which skills are more necessary for the CIO to successfully perform his job. The research also examined how the CIO views his role in the organization, his communication skills as well as his skills in general. A semi-structured in-depth interview with CIOs from 10 Croatian private and public organizations will explore their position, status, work experience in ICT, role, and communication skills. The *in-depth interview protocol* is based on sources that deal with modern communication skills and the role of CIOs in the world and Croatia. Some questions arise based on the content of research papers that are closely related to the topic of this research [9, 28, 31–34]. The protocol consists of four parts:

1. General information about the CIO
2. Role and position in the organization
3. Communication role and
4. His skills

Although other research papers recommend recording conversations, this was not done in any of the interviews due to challenges in accepting interviewing as a method in general. Several quantitative questions were asked during the in-depth interview with CIOs, and quantitative results were obtained. These are as follows:

1. Data on how much time they spend on certain types of common tasks.
2. How much time they spend on collaboration and communication, how much on meetings, and how much on office work.
3. The ratio between written and oral communication in everyday work.
4. The way they educate themselves for their role.

The sample was first divided into private and public organizations, and then into organizations where CIOs have more and organizations where CIOs have less work experience in their role as CIOs. CIOs with less than 12 years of experience were assigned to the group with less work experience and those with 12 or more years of experience were assigned to the group with more work experience.

4 Results and Discussion

4.1 General Information About the CIOs

According to the collected data, it was determined that all ten CIOs have an excellent educational background, seven of them have a university degree, two of them have a PhD, and one has a master of science degree. According to their age, four of them are between 30 and 40 years old, four are between 40 and 50 years old, and of the remaining two CIOs, one is over 50 and the other over 60 years old. According to the title of the function they perform, there are four *CIOs*, three *IT managers*, two *IT directors,* and one with two functions – *Head of IT support and financial controlling*. All of the interviewed CIOs are married and have children.

4.2 The Role and Position of the Interviewed CIOs

The CIOs of this research are predominantly in high hierarchical positions. They are either members of the board of directors or extended management, or they report to the executive director. Only three of them report to the CFO *chief financial officer*. This means that the organizations in which they work, in most cases, value their role correctly. It is important to note that the CIOs from three organizations, in which the management has adequate ICT knowledge to understand ICT, are involved in the strategic decisions of their organization, while in the other seven organizations, where the management is not adequately ICT literate, only one CIO participates in making strategic decisions for the organization. To the question about how they assess the importance of their role in the organization and whether it is clear to everyone, seven of them answered that their role in the organization is very important, two answered that it is partially important, and only one said that it is not important at all.

4.3 The Communication Role of the Interviewed CIOs

The results of this research indicate a big difference between global statistics and the results in the part about verbal and written communication. On average, the CIOs said that both types of communication are equally represented in their daily work, more precisely 48% of their time is spent on oral communication and 52% on written communication. Interestingly, the percentage difference in public organizations is much greater in favor of written communication, 58% versus 42%. The reason for this, as was later commented on during the in-depth interview, lies in the fact that in public organizations, there is more insistence on written evidence, and communication is more formal. The legal regulation that favors this type of communication should not be ignored either. In private organizations, communication is faster because of the context in which it takes place. The competitive environment does not leave much room for long delays because it is necessary to react quickly and there is no time for formal correspondence and waiting for responses. All respondents confirmed that the ICT skills of ICT users and management members present a problem for the CIOs from this research. They emphasize the need for mutual adaptation in order to *speak the same language*.

4.4 The Most Important Skills of the Interviewed CIOs

The CIOs from this research are aware that a person who strives for success cannot be a so-called classic manager and that in addition to all the important managerial knowledge, a very wide range of responsibilities and skills should be covered. According to them, communication skills should stand out the most because you need to be an excellent communicator to have good and successful relationships [20]. The art of communication is also important for building good relations between all interested parties. CIOs are required to have a vision of how to build the success of the organization as well as the ability to communicate ICT topics in the *language of business* [35]. This was confirmed by almost all of the managers who were questioned in the in-depth interview. It is important to note that in their list of the most important factors, nine out of ten respondents had one, and some even two, factors that could be classified as communication skills [4]. Despite the different gradations of the most important factors, all CIOs are very aware of the importance of communication skills for personal and organizational success. A striking fact in these results is that only one CIO, from a public organization, thinks that continuous education is an important criterion for success. All of them mentioned the importance of being familiar with business aspects and processes as well as having technical knowledge and proficiency. CIOs are aware that skills in the field of finance are very much necessary because without it there is no management and control, and this awareness is the result of their long, successful careers. The majority of CIOs in this survey believe that formal education is not crucial for success, but they do think that education provides them with broad knowledge and serves as a good basis for a better-paid job. Formal education at universities is facing problems because a parallel system of education has emerged that sometimes creates better employees for modern business needs. Based on the research conducted by Trigo et al. among Portuguese CIOs, the CIOs in this study also rated the importance of the seven most important skills on a seven-point Likert scale. Trigo et al. examined the importance of the same skills among CIOs (54 respondents), but they used a five-point Likert scale [28], as shown in Fig. 1. The importance of certain skills differs significantly from the original research conducted among Portuguese CIOs [28] and the CIOs of this study. In this research, *technical proficiency, the ability to follow technological innovations,* and *negotiation skills* are considered the most important. The CIOs in the research conducted by Trigo et al. [28] assessed

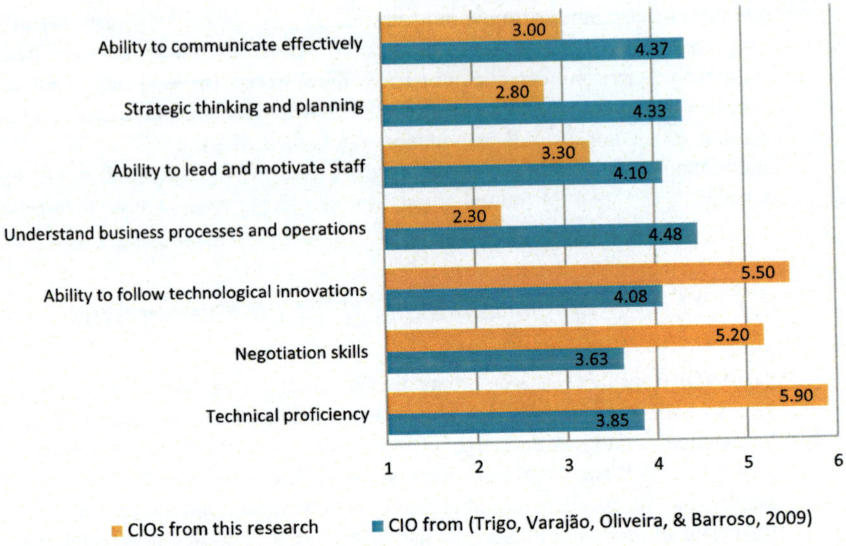

Fig. 1 Descriptive statistics – arithmetic mean of answers given by Croatian and foreign CIOs to questions about the importance of the seven listed skills

the importance of those three skills with the lowest scores, but they assessed *the understanding of business processes and operations,* ***the ability to communicate effectively****, and the skill of strategic thinking and planning* as the most important. A few years earlier, using the same set of questions in Germany, Sojer et al. received results in which these same three skills were ranked best, but here the *ability to communicate effectively* was classified as the most important [36]. The CIOs of this research surprisingly ranked the ability to communicate effectively as the fifth out of seven possible levels of importance. The congruence in the answers was also evaluated using the Pearson correlation coefficient. Assessments of the importance of CIOs' skills from this study are rather negatively related to assessments of CIOs from the Trigo et al.'s research [28] ($r = -0.841$; $p = 0.18$).

Such a high correlation points to important differences regarding the CIO's role and skills between Croatian and foreign CIOs. What raises even more questions are the differences in the results between this and the large American study. A 2011 ECAR[1] survey of 3400 CIOs from 1000 US colleges and universities shows that communication skills are the most important. An interesting fact is that in this survey, the importance of technical proficiency is in last place with only 31% [37]. Even though this was a study conducted among university CIOs, it is important to point out the global trend when it comes to the importance of communication skills. Such results of our CIOs are quite unexpected, considering global trends and

[1] Educase center for analysis and research.

statistics, but they coincide with previous research conducted by Mađarić et al. about the roles and responsibilities of CIOs in Croatia [34].

4.5 Limitations of This Research

The main shortcoming of this research is manifested in the insufficient number of respondents, and consequently also in the impossibility of applying the research results to a specific business or public sector area and presenting the state of the profession with the specifics that each of them has. Choosing the CIOs and getting their consent to participate in the research was also a relevant factor. Many CIOs were contacted. Some did not respond at all, even after several inquiries. Some of them declined due to organizational politics, lack of time, and similar reasons. For one to remain at the top of any profession, continuous formal education is necessary. The same situation applies to CIOs, even though only one out of ten answers emphasized the importance of continuous formal learning. There is a possibility that the CIOs misunderstood the scope and essence of this question, or perhaps the question was phrased inappropriately.

4.6 Some Suggestions for Future Research

First of all, it would be advisable to conduct research on a larger sample of CIOs in order to get a clearer picture. Accordingly, individual business areas and activities of the examined organizations should be differentiated. We are aware of the fact that organizations in the public sector often have different priorities than those in the private sector, as this research has shown [38]. It would be good to directly investigate the difference between the private and the so-called public sector in the context of the communication role because then one could expect an even clearer picture of the status and skills of CIOs depending on the environment in which they work. By processing the data collected in this research, the need for additional research on communication priorities has been determined. It is necessary to examine how these priorities are seen by the management on the one hand, and by the CIOs on the other hand, considering the communication channel that they have mutually established. The research was conducted in organizations that still own the entire or at least the majority of the organization's ICT systems. It would be desirable to investigate the change in the role and required skills of CIOs in companies that have outsourced their ICT system. In such companies, the maintenance of the infrastructure and most of the services are transferred to another entity, which changes the status, responsibilities, and required skills of the CIO. We should investigate and try to find an explanation for why communication skills are not on the list of priorities when it comes to Croatian CIOs when these are clearly the most valued skills in the world.

This research was conducted in Croatia with a small number of participants and is therefore limited, but it is advisable to conduct it with the same set of questions in several European countries. It should be determined whether European CIOs, except those from Portugal and Germany, are indeed more oriented toward the acquisition of technical knowledge or whether this is only the case in Croatia.

5 Conclusion

If he wants to be an active member of society and a creator of new values, the CIO is expected to be an excellent communicator, in addition to all other much-needed skills. It is imperative to overcome the decades-old paradigm according to which the CIO is only a technologist who works in his "own world." The new CIO is focusing on business and becoming an important partner of the board in the business decision-making process. And in order to achieve greater influence, he should become a part of the board and directly participate in making business decisions. For his new role, he needs excellent communication skills and business knowledge, because he must communicate more and more in the language of business and use less technical terms. In concrete situations, this would mean that the CIO should effectively communicate business strategies, implement ICT plans, and effectively manage and monitor operational progress [30]. That is why the Croatian CIO needs to be more aware of the importance of his role in the organization and his communication role. Despite a large number of relevant international research papers that put communication skills first, this research confirmed the well-known thesis that Croatian CIOs think that technical proficiency is still more important. It is necessary to work on this topic so that CIOs become aware of the importance of communication skills for success. Technical proficiency and the ability to follow technological innovations are important, but not crucial. Because without clear and effective communication [36], it is not possible to guarantee success in the areas that the CIO deals with.

Given the fact that we are living in a globalized world where different business cultures are coming together, universities and higher education institutions, in general, must be aware of the new paradigm according to which we need to raise awareness about the importance of communication skills for the success of an individual, and therefore of the society as a whole. It would be advisable to incorporate a set of courses in the curricula of technical universities that most CIOs attend that would help develop their communication skills and prepare them for other business challenges they may encounter.

References

1. J.-H. Wu, Y.-C. Chen, J. Chang, Critical IS professional activities and skills/knowledge. A perspective of IS managers. Comput. Hum. Behav. J. **23**(6), 2945–2965 (2007)

2. A. Hollman, S. Bickford, J. Lear, Communication processes of information technology executives in higher education. J. End User Comput. **30**, 2 (2018)
3. J. Peppard, C. Edwards, R. Lambert, Clarifying the ambiguous role of the CIO. MIS Q. **10**, 1 (2011)
4. C. Stephens, T. Loughman, The CIO's chief concern: Communication. Inf. Manag. **27**(2), 129–137 (1994)
5. R.D. Banker et al., CIO reporting structure, strategic positioning, and firm performance. MIS Q. **35**(2), 487–504 (2011)
6. J. Luftman, *IT-Business Alignment and the Evolving Role of the CIO and IT: 2017–18*. ResearchGate.net. [Online] (2018). https://www.researchgate.net/profile/Rajeev-Dwivedi-Phd/publication/283214076_Influential_IT_management_trends_an_international_study/links/5e5639bc92851cefa1c494a0/Influential-IT-management-trends-an-international-study.pdf
7. B. Ives, M.H. Olson, Manager or technician? The nature of the information systems manager's job. MIS Q. **5**(4), 49–63 (1981)
8. W.R. Synnott, W.H. Gruber, *Information Resource Management: Opportunities and Strategies for the 1980s* (John Wiley & Sons, New York, 1981)
9. M. Chun, J. Mooney, CIO roles and responsibilities: Twenty-five years of evolution and change. Inf. Manag. **46**(6), 323–334 (2009)
10. M. Carter, V. Grover, J.B. Thatcher, The emerging CIO role of business technology strategist. MIS Q. Exec. **10**(1), 19–29 (2011)
11. W.R. Synnott, *Putting a CIO in Charge*. Institutional Investor, December 1987, pp. 47–48
12. H.G. Enns, D.B. McFarlin, S.L. Huff, How CIOs can effectively use influence behaviors. MIS Q. Exec. **6**, 1 (2007)
13. S.C. Certo, S. Trevis, *Moderni menadžment*, 10th edn. (Mate, Zagreb, 2008)
14. Webster Inc., *An Encyclopædia Britannica Company*. Dictionary. [Online] (2012). www.merriam-webster.com/dictionary/communication
15. Biblija, Kršćanska sadašnjost, Zagreb, 1991
16. H. Mintzberg, The manager's job: Folklore and fact. Harv. Bus. Rev., 49–61 (1975)
17. J.D. Couger, R. Zawacki, E.B. Oppermann, Motivation levels of MIS managers versus those of their employees. MIS Q. **3**(3), 47–56 (1979)
18. J.F. Rockart, *The Changing Role of the Information Systems Executive: A Critical Success Factors Perspective* (Sloan School of Management, MIT, Cambridge, 1982), pp. 3–11
19. R.I. Benjamin, C. Dickinson, J.F. Rockart, Changing role of the corporate information systems officer. MIS Q. **9**, 177–188 (1985)
20. D.F. Feeny, B.R. Edwards, K.M. Simpson, Understanding the CEO/CIO relationship. MIS Q. **16**(4), 435–448 (1992)
21. M.J. Earl, D.F. Feeny, Is your CIO adding value? McKinsey Q. **2**, 144–161 (1995)
22. J.T. Karlsen, P. Gottschalk, E.S. Andersen, Information technology management roles: A comparison of IT executives and IT project managers, in *Proceedings of the 35th Hawaii International Conference on System Sciences*, (IEEE, 2002), pp. 1–8
23. J.W. Weiss, D.J. Anderson, CIOs and IT professionals as change agents, risk and stakeholder managers: A field study. Eng. Manag. J. **16**(2), 13–18 (2004)
24. A.M. Johnson, A.L. Lederer, The effect of communication frequency and channel richness on the convergence between chief executive and chief information officers. J. Manag. Inf. Syst. **22**(2), 227–252 (2005)
25. D.H. Smaltz, V. Sambamurthy, R. Agarwal, The antecedents of CIO role effectiveness in organizations: An empirical study in the healthcare sector. IEEE Trans. Eng. Manag. **53**(2), 207–222 (2006)
26. E. Karahanna, R.T. Watson, Information systems leadership. IEEE Trans. Eng. Manag. **53**, 171–176 (2006)
27. D.Q. Chen, D.S. Preston, Understanding CIO role effectiveness: The antecedents and consequents, in *HICSS '07 Proceedings of the 40th Annual Hawaii International Conference on System Sciences*, (IEEE Computer Society, Washington, DC, 2007)

28. A. Trigo et al., Chief Information Officer's activities and skills in Portuguese large companies. Commun. IBIMA J. **10**, 64–71 (2009)
29. S. Mitra, V. Sambamurthy, G. Westerman, Measuring IT performance and communicating value. MIS Q. Exec. **10**(1), 47–59 (2011)
30. C. Cano, L. Fernández-Sanz, S. Misra, Featuring CIO: Roles, skills and soft skills. Int. J. Hum. Cap. Inf. Technol. Prof. **4**(1), 12 (2013)
31. C.S. Stephens et al., Executive or functional manager?: The nature of the CIO's job. MIS Q. **16**(4), 449–467 (1992)
32. P. Weill, S.L. Woerner, The future of the CIO in a digital economy. MIS Q. Exec. **12**(2), 65–75 (2013)
33. P. Gottschalk, N.J. Taylor, Strategic management of IS/IT functions: The role of the CIO, in *Proceedings of the 33rd Hawaii International Conference on System Sciences*, (IEEE, 2000)
34. M. Mađarić, Z. Krakar, V. Lovrek, Roles and responsibilities of modern CIO – Status and analysis of trends in Croatia, in *Central European Conference on Information and Intelligent Systems, CECIIS – 2009*, (Faculty of Organization and Informatics, Varaždin, 2009), pp. 167–176
35. M. Broadbent, E. Kitzis, *The New CIO Leader: Setting the Agenda and Delivering Results* (Harvard Business School Press, Boston, 2004)
36. M. Sojer, C. Schläger, C. Locher, The CIO – Hype, science and reality, in *14th European Conference on Information Systems Proceedings*, (European Conference on Information Systems, Göteborg, 2006)
37. P. Arroway et al., *The Higher Education CIO, Portrait of Today, Landscape of Tomorrow* (EDUCASE, Center for Applied Research, Washington, 2011)
38. W. Noonpakdee, A. Phothichai, T. Khunkornsiri, A. Nuntree, CIO competency in digital era: A comparative study between government organizations and private enterprises, in *2020 IEEE 7th International Conference on Industrial Engineering and Applications (ICIEA)*, (IEEE, 2020), pp. 948–952

Part III
Sustainable Industrial Development and Industry 4.0

Concept of Mixed Reality Application Design for Technical Solutions

Jozef Husár, Stella Hrehová, Lucia Knapčíková, and Justyna Trojanowska

1 Introduction

In today's modern technological environment, there is a constantly growing need for effective and innovative ways of solving technical problems. Mixed reality (MR) is becoming one of the revolutionary technologies that brings a new way of interaction between the virtual and real world. The use of mixed reality in technical solutions offers countless possibilities in areas such as maintenance, repair, diagnostics, and visualization of technical elements [1]. The aim of this article is to present a design concept for a mixed reality application that would help engineers and technicians in the area of technical solutions improve their work and achieve better results. This application provides users with an intuitive user interface that allows easy interaction with virtual and real objects. By detecting and tracking physical objects, the app could connect virtual information with the real world, providing users with important technical information, such as repair instructions, manuals, or spare parts, through virtual layers that would be displayed on the mixed reality device [2]. The design of the application includes the possibility of displaying 3D technical elements models and devices, which allows users to more accurately examine and visualize these objects. Interactive multiuser collaborations can be an extension of this application, allowing engineers to share their insights and advice through

J. Husár (✉) · S. Hrehová · L. Knapčíková
Technical University of Košice, Faculty of Manufacturing Technologies, Department of Industrial Engineering and Informatics, Prešov, Slovak Republic
e-mail: jozef.husar@tuke.sk; stella.hrehova@tuke.sk; rebekalucia.knapcikova@tuke.sk

J. Trojanowska
Poznan University of Technology, Faculty of Mechanical Engineering, Department of Production Engineering, Poznan, Poland
e-mail: justyna.trojanowska@put.poznan.pl

L. Knapčíková, D. Peraković (eds.), *8th EAI International Conference on Management of Manufacturing Systems*, EAI/Springer Innovations in Communication and Computing, https://doi.org/10.1007/978-3-031-53161-3_10

mixed reality. This would improve collaboration and communication among teams in different locations. The app also includes tools to simulate and train technical solutions, which would allow users to try out different scenarios, test their solutions, and get feedback directly through mixed reality [3].

1.1 Mixed Reality

Mixed reality is a relatively new technology that combines virtual reality and augmented reality, where virtual objects interact with real objects in physical space. Mixed reality represents the next wave in computing technology, where computers and smartphones are replaced by holographic smart glasses. It is becoming mainstream for consumers, businesses, and freeing us from screen-bound experiences by offering instinctive interactions with data in our living spaces and with our friends [4].

Paul Milgram and Fumio Kishino first mentioned the term "mixed reality" in 1994 to describe the continuum between completely real and completely virtual environments [5, 6]. Mixed reality is a combination of virtual reality and augmented reality, and is an environment in which real and virtual subjects and objects interact in real time, allowing users to interact with both real and virtual components. If we want to experience mixed reality, we need a headset with a transparent lens or a camera that allows the user to keep an eye on the real world [7].

2 Mixed Reality Tools

This part is devoted to a detailed software and hardware description, which we used to create the application based on the mixed reality principle. Mixed reality combines the physical and digital worlds and creates a new and immersive experience for users. It offers a unique way to interact with virtual objects in the real world, minimizing the boundaries between what is real and what is not. For this activity, it is therefore necessary to define the basic requirements for mixed reality tools [8].

2.1 Unity

Mixed reality can also be used with the help of Unity software, in which we can create our proposed application. Unity is a cross-platform game engine developed by Unity Technologies, first introduced and announced in June 2005 at Apple Inc.'s Worldwide Developers Conference as an exclusive game engine for Mac OS X. As of 2018, the software has been expanded to support more than 25 platforms. Unity is a tool used by a large number of developers to create games and their creations. Unity software is powerful, easy to use, and accessible to everyone [9, 10].

Unity provides a powerful cross-platform development environment. As a game engine, Unity software provides many important built-in features that make games work. This means things like object physics, 3D rendering, and collision detection.

The Unity software becomes even more useful because it also includes a thriving "Asset Store," which is the basic place where developers can upload their creations and make them available to the community. This means that game developers can focus on what matters, like designing unique and fun experiences, because they only need to choose the materials they can choose from this store to create them. Through the Unity Asset Store, creators can develop and sell user-created works from other game creators. These include 3D and 2D assets and environments that developers can buy and sell. The Unity Asset Store was launched in 2010 [11].

The Unity engine provides the Unity editor with the main scripting API of the C# programming language, as well as drag-and-drop functionality, both in the form of plugins and in the game itself. Before C# became the main programming language used for this software, it was supported by Boo, which was removed with the release of Unity 5, and a version of JavaScript called UnityScript, which was removed in August 2017 after the release of Unity 2017 [12].

Mixed reality is an area where Unity excels, as it provides a range of tools and features to create immersive mixed reality experiences. Unity enables developers to create mixed reality experiences by leveraging its support for a variety of hardware platforms, including Microsoft HoloLens, Magic Leap, and other AR/VR glasses. Developers can use Unity's scripting tools and visual editor to create interactive objects and environments that can be overlaid onto the user's physical environment or integrated with real-world objects [13].

The Unity Mixed Reality Toolkit (MRTK) provides a set of features and components that make it easy to create mixed reality experiences in Unity (Fig. 1). MRTK includes features such as hand tracking, spatial mapping, and voice recognition that can be used to create immersive and interactive mixed reality applications [14].

Additionally, Unity supports a wide range of AR/VR devices and platforms, making it easy for developers to create cross-platform mixed reality applications. With Unity, developers can create mixed reality experiences that are compatible with a variety of devices and operating systems, including iOS, Android, Windows, and more [15].

The Mixed Reality Toolkit (MRTK) is a set of tools and components provided by Microsoft that facilitate the development of mixed reality applications in Unity. MRTK provides a range of pre-built components and features that can be used to create interactive and immersive mixed reality experiences [16, 17].

Below are some of the key features provided by the MRTK toolkit:

- Hand sensing: provides support for hand tracking, allowing developers to create experiences where users can interact with virtual objects using their hands.

Fig. 1 MRTK toolkit for using mixed reality

- Spatial mapping: includes components that allow developers to create spatial maps of the user's environment that can be used to create mixed reality experiences that are aware of the user's surroundings.
- Input management: provides a framework for managing input from a variety of devices, including hand controllers, voice recognition, and more.
- Interactive objects: it also has pre-built components for creating interactive objects such as buttons, sliders, and switches.
- Visual effects: delivers a range of visual effects such as particle systems and post-processing effects that can be used to enhance the immersive quality of mixed reality experiences.

In addition to these features, MRTK also provides a variety of tools and resources for developing mixed reality applications, including documentation, tutorials, and sample projects. MRTK is open source and available on GitHub, allowing developers to contribute to its development and customize its features to suit their specific needs [18, 19].

2.2 *Headset Microsoft HoloLens 2*

Microsoft HoloLens 2 is a mixed reality headset designed and manufactured by Microsoft. It is a wearable device that allows users to interact with digital holograms in their physical environment. HoloLens 2 is designed for use in various industries such as healthcare, manufacturing, education, and more (Fig. 2). As this is a complex device, it is necessary to list the important components that work together to provide a fully valued mixed reality experience [20].

Fig. 2 Microsoft HoloLens 2

3 Materials and Methods

In this section, the step-by-step procedure for creating an application usable for the presentation of the MR option is defined. The task of the application is the possibility of interacting with objects using the Microsoft HoloLens 2 headset, in which these objects can be enlarged, reduced and rotated with them, as well as the placement of an object or an industrial machine in the space, for example, a lathe, which would provide us with information about how what the object might look like in the physical world [21, 22].

The first step we take after launching the Unity software is to configure the build settings (Build Settings). In these settings, we choose the target platform, which is the Universal Windows Platform (UWP) and adjust the Architecture option to ARM 64-bit, the other options are not changed. ARM64 is a type of processor architecture based on the ARMv8-A architecture, which is a 64-bit architecture developed by ARM Holdings. The ARM64 architecture is designed to provide better performance and energy efficiency compared to its 32-bit predecessor ARMv7. It is commonly used in smartphones, tablets, and other mobile devices, as well as in servers, embedded systems, and other applications that require high performance with low power consumption [23].

After successfully changing the assembly settings, it is necessary to install the mixed reality toolkit – MRTK, from the website Microsoft.com/download and extract the downloaded file to the computer.

After installing the tools for mixed reality, it is necessary to choose a storage location, which will be a folder. We will add the "Discover Features" feature, where we can choose from several mixed reality features as needed for the project. To create our project, it is necessary to check the option "Mixed Reality Toolkit Foundation" version 2.8.0 and in the "Platform Support" folder "Mixed Reality

OpenXR Plugin" version 1.8.0. After choosing the functions for the project, click the "Get Features" button and they will be installed.

After installing the mixed reality functions, we click the "Validate" button (Verify) if we did not encounter any errors during the package features installation. After this step, we have successfully installed and embedded the mixed reality toolkit in our project. In the Unity software, the configurator of the mixed reality tool package will open, which will guarantee that the MRTK package is ready for use without any problems. We import "Unity OpenXR plugin." OpenXR is a royalty-free, freely distributable add-on developed by Khronos that aims to simplify virtual, augmented, and mixed reality development by enabling developers to seamlessly target a wide range of VR/AR/MR devices [24].

With OpenXR, it is possible to build tools that target holographic devices like HoloLens 2 as well as immersive virtual reality devices. OpenXR allows writing code that is then portable across a wide range of hardware platforms.

Subsequently, the project settings (Project Settings) will be displayed with the "XR Plug-in Management" box open. In it, we select the Universal Windows Platform setup and check the "OpenXR" and "Microsoft HoloLens feature group" options so that they are started.

We set Open XR and "Depth Submission Mode" to "Depth 16 Bit". In the "Interaction Profiles" box, select the "Microsoft Hand Interaction Profile" option. Thanks to this option, we will be able to interact with objects in the application using our hands. We will prepare the user interface and add "TMP Essentials" (TextMesh Pro) to the project. TextMesh Pro provides better control over text formatting and layout. Using TextMesh Pro in your projects will allow you to break out of the usual set of fonts used by Unity software into the realm of unlimited possibilities. It includes features such as spaces between characters, words, lines and paragraphs, text alignment, links, more than 30 labels with formatted text, support for multiple fonts, advanced rendering of text using custom shaders [25].

3.1 Import 3D Models

We can add our objects in different ways, either using the Asset Store or through any CAD software. The objects added by us were through the Asset Store were edited and inserted into the project in the Unity software [26, 27]. The selected models are a lathe and a mill (Fig. 3).

In the package manager, we click on the items of the machine models that we added through the store. We will perform the same procedure for both machine models (Fig. 4).

Newly inserted models from either the store or CAD software are not defined at all in Unity and each model we insert into the software has its own size dimensions and does not have a component for collisions, as 3D models that are ready to use in advance in Unity software [28]. Therefore, it is necessary to edit them in the Inspector window, where we will edit their dimensions. We apply the same

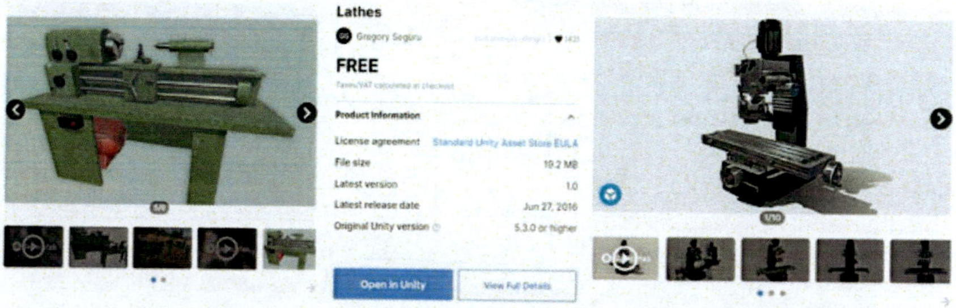

Fig. 3 Imported models from the asset store

Fig. 4 Model insertion

procedure to define the lathe model and the machine model for milling, except that we will not use the "NearInteractionGrabbable" script, because we will also perform the interaction of these models from a distance (Fig. 5).

This is how we create an MR application. Even though we have an app built in Unity software, it is not possible to run it using the headset. Before testing the application through the headset, we need to enable developer mode in the Windows operating system settings in order to connect the headset to the computer, as this feature is normally disabled. We apply the same procedure to the headset. The developer mode setting enables debugging and other development options, such as writing software in Microsoft Visual Studio. The next step is to pair the headset with the computer. Under the developer mode in the operating system of the Windows Holographic headset, we click the "Pair" button and a window pops up with a 6-digit PIN code that we enter in the computer's device settings. After entering it, we will

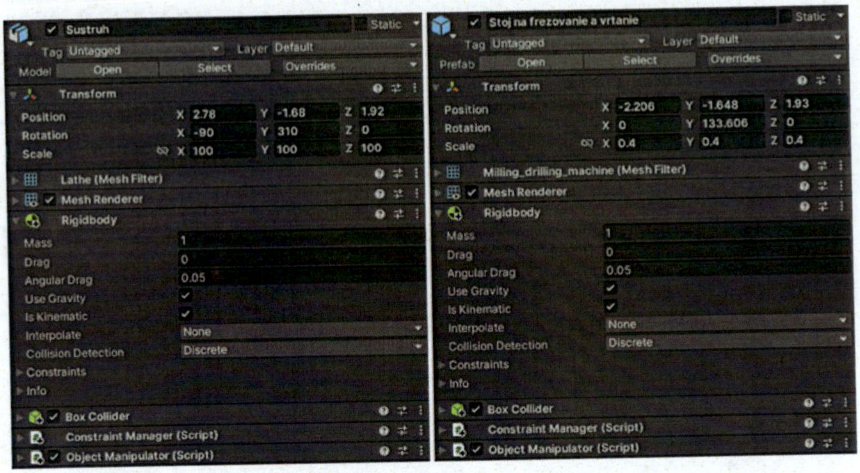

Fig. 5 Defining components for machine models

have the Microsoft HoloLens 2 headset successfully paired with the computer. We start the application using the headset using the Microsoft Visual Studio program, by double-clicking the MRTK.sln box in the folder where we have saved the "Build" of the application from the Unity software.

4 Results

We can test the application created by us at any time in the Unity software in the Game window (Fig. 6). In the application, we move with the W, A, S, D keys by holding the "Space" (right hand) or "Shift" (left hand) keys and at the same time by pressing and dragging the left mouse button we can move objects and if we want to rotate objects, we do it by pressing the "Space" or "Shift" keys, the left "Ctrl" and pressing and dragging the left mouse button.

The created application allows us to work with the devices created by us. In general, we can create a generalized algorithm (Fig. 7) described in detail in chapter "Sustainability and Circular Economy Through Application and Processing of Recycled Materials into Additive Manufacturing".

After generalization, we can characterize the algorithm for creating a mixed reality application with eight steps:

1. Step: Definition of application goals

   Identify and define the goals and requirements of your mixed reality application. Consider whether it is for a specific area, such as training, design, collaboration, or entertainment.

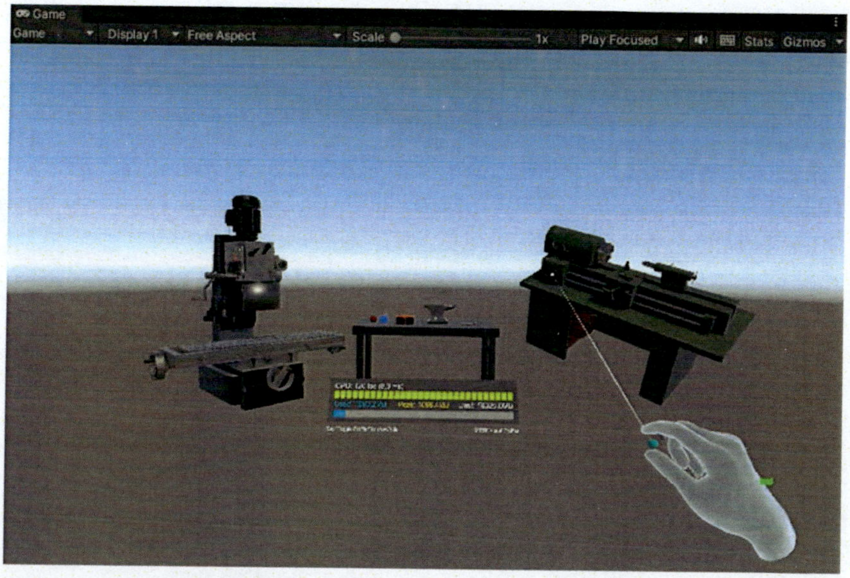

Fig. 6 Unity test user interface

2. Step: Choosing the right platform

   Choose the right platform for your mixed reality application. You can consider platforms such as Microsoft HoloLens, Magic Leap, Oculus Rift, or creating applications for mobile devices with support for ARKit (iOS) or ARCore (Android).
3. Step: 3D models creation and virtual content

   Create or acquire 3D models and virtual content, which will be integrated into the application. This can include objects, characters, interactive elements and environments.
4. Step: Implementation of location tracking and marker recognition

   Integrate location tracking and marker recognition into your app to ensure virtual content is correctly positioned and included in the real world.
5. Step: Implementation of interaction with virtual content

   Provide opportunities to interact with virtual content. This may include gestures, voice control, touch control, or control using a control device.
6. Step: User interface creating

   Design and implement a user interface that will allow users to navigate and control the mixed reality application. Consider intuitive controls, information display methods, and visual presentation of virtual content.
7. Step: Testing and debugging

   Test the app and debug features to ensure it works properly. Try different scenarios and verify whether the application meets the requirements and achieves the set goals.

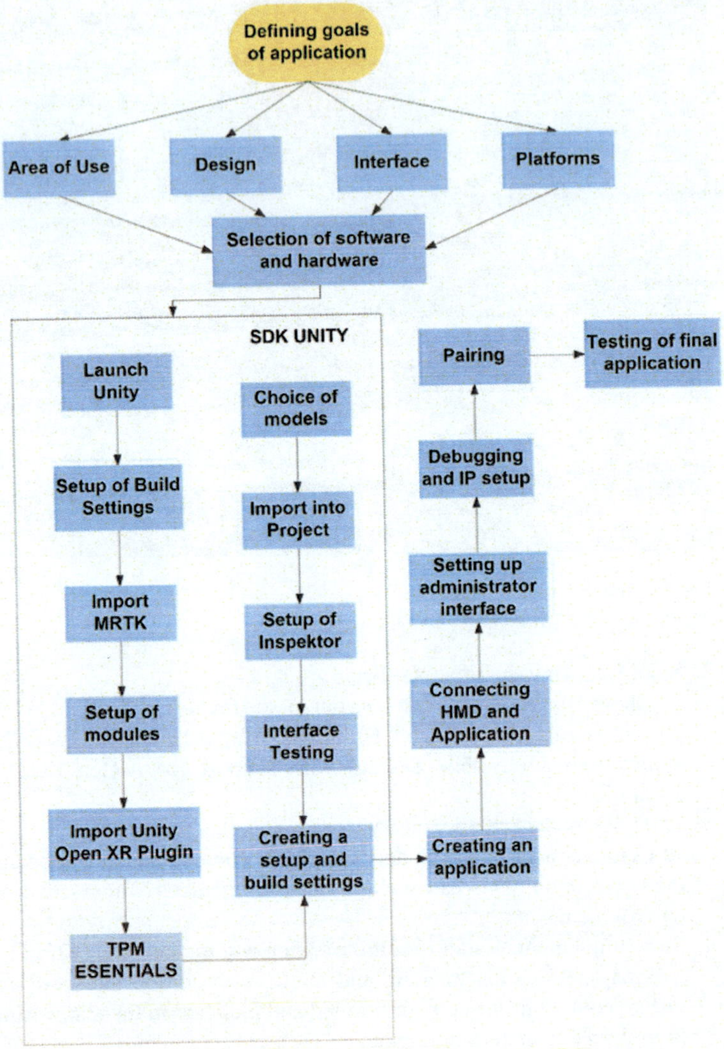

Fig. 7 MR application creation algorithm

8. Step: Distribution and commissioning
   Prepare the application for distribution and commissioning. You can publish it in the respective stores with the Apple Store and Google Play applications.

These generalized steps apply to the universal creation of mixed reality applications.

5 Conclusion

At a time of constant manufacturing industry development, it is believed that MR technology is the basis for realizing the vision of Industry 4.0. Mixed reality (MR) has created new ways of interaction for users and computer systems. In general, we can define several main benefits of MR in technical practice:

- Realistic visualization: MR enables accurate and realistic virtual object visualization in the real environment, which helps in the design and visualization of technical solutions.
- Interactive simulations: MR enables the creation of interactive simulations where engineers and technicians can test and optimize different scenarios and configurations in real time.
- Improved design and prototyping: Thanks to MR, engineers can create virtual prototypes and test them in a real environment, enabling faster and more efficient iterations in the design of technical solutions.
- Training and education: MR provides an interactive and realistic environment for training technicians and employees. It can be used for simulation and training in various technical fields.
- Improved collaboration and communication: MR enables technicians and engineers to collaborate in real time and share information and visual data within projects. This contributes to increasing the efficiency and accuracy of the team's work.
- Improved maintenance and repairs: With the help of MR, it is possible to display maintenance and repair instructions directly on the equipment in real time. Technicians thus have better access to information and can perform maintenance and repairs more efficiently.
- Process optimization: MR can be used to monitor and analyze processes in real time. It helps to identify weak points and improve the efficiency of technical processes.
- Increased safety: MR can be used to create virtual safety training and simulations, which helps to improve safety procedures and minimize risks.
- Personalized solutions: MR allows technicians and engineers to create personalized solutions and environments for different purposes, increasing efficiency and comfort at work.

Our presented article has the task to bring the possibilities of displaying two selected machines in MR. It provides a manual (step-by-step post) and a generalized algorithm for creating an MR application. This application allows working with virtual objects. In the holographic interface, it is possible to disassemble the selected machines into components and interact with them in space as desired. The article is suitable background material for designers, and students for education, but also for small and medium-sized enterprises (SMEs) that can train their employees in this way. In the future, the goal of the authors of the paper is to create an MR application that will work with model libraries and download data from an online database.

Acknowledgments This work was supported by the projects VEGA 1/0268/22, KEGA 038TUKE-4/2022 granted by the Ministry of Education, Science, Research and Sport of the Slovak Republic.

References

1. O. Blimber, *Spatial Augmented Reality: Merging Real and Virtual Worlds* (A K Peters, Wellesley, 2005)
2. J. Husár, L. Knapčíková, Possibilities of using augmented reality in warehouse management: A study. Acta Logist. **8**(2), 133–139 (2021). https://doi.org/10.1007/10.22306/al.v8i2.212
3. T. Mazuryk, M. Gervautz, *Virtual Reality-History, Applications, Technology and Future* (TU Wien University, Vienna, Austria, 1996)
4. O. Hansson, O. Nadum, Model based operator instructions using AR-technology. M.S. thesis, Department of Industrial and Materials Science, Chalmers University of Technology, Gothenburg, Sweden, 2019. https://hdl.handle.net/20.500.12380/300753. Last accessed 2022/11/04
5. P. Milgram, H. Takemura, A. Utsumi, F. Kishino, Augmented reality: A class of displays on the reality-virtuality continuum, in *Telemanipulator and Telepresence Technologies*, ed. by H. Das, (SPIE, Bellingham, 1995), pp. 282–292. https://doi.org/10.1117/12.197321
6. P. Milgram, F. Kishino, A taxonomy of Mixed Reality visual displays. IEICE Trans. Inf. Syst. **E77-D**(12), 1321–1329 (1994)
7. C. Forrest, Infographic: The history of AR and VR, and what the future holds, https://www.techrepublic.com/article/infographic-the-history-of-ar-and-vr-and-what-the-future-holds/. Last accessed 2022/11/04
8. Y. Zinchenko, P. Khoroshikh, A. Sergievich, et al., Virtual reality is more efficient in learning human heart anatomy especially for subjects with low baseline knowledge. New Ideas Psychol. **59**, 100786 (2020). https://doi.org/10.1016/j.newideapsych.2020.100786
9. A. Mahmutovic, A. Andreasson, L. Söderström, Effektivisering av plockprocessen i lager med hjälp AV AR. Bachelor thesis, Department of Industrial and Materials Science, Chalmers University of Technology, Gothenburg, Sweden, 2018. https://hdl.handle.net/20.500.12380/255328. Last accessed 2022/11/03
10. V. Kaushal, Exploratory study: Implementation and applications of extended reality. Master's thesis, Faculty of Science and Technology, University of Stavanger, Department of Mechanical and Structural Engineering and Materials Science, 2019
11. A. Heinzel, S. Azhar, A. Nadeem, Uses of augmented reality technology during construction phase, in 9th International Conference of Construction in the Twenty-first Century (CITC9), 5–7 March 2017, Dubai, UAE (2017)
12. J. Kaščak, J. Husár, L. Knapčíková, J. Trojanowska, V. Ivanov, Conceptual use of augmented reality in the maintenance of manufacturing facilities, in *Advances in Manufacturing III. MANUFACTURING 2022*, Lecture notes in mechanical engineering, ed. by J. Trojanowska, A. Kujawińska, J. Machado, I. Pavlenko, (Springer, Cham, 2022). https://doi.org/10.1007/978-3-030-99310-8_19
13. Unity. https://unity.com/. Last accessed 2022/11/04
14. J. Kascak, M. Teliskova, J. Torok, et al., Implementation of augmented reality into the training and educational process in order to support spatial perception in technical documentation, in Proceeding of the IEEE 6th International Conference on Industrial Engineering and Applications (ICIEA), 26–29 April 2019, Tokyo, Japan (2019), pp. 583–587, https://doi.org/10.1109/IEA.2019.8715120
15. H. Dhiman, S. Buttner, C. Rocker, R. Reisch, Handling work complexity with AR/deep learning, in Proceedings of the 31st Australian Conference on Human-Computer-Interaction (OZCHI'19), 2–5 December 2019, New York, USA (2020), pp. 518–522, https://doi.org/10.1145/3369457.3370919

16. F. Lamberti, F. Manuri, A. Sanna, G. Paravati, P. Pezzolla, P. Montuschi, Challenges, opportunities, and future trends of emerging techniques for augmented reality-based maintenance. IEEE Trans. Emerg. Top. Comput. **2**(4), 411–421 (2014). https://doi.org/10.1109/TETC.2014.2368833
17. X.W. Chen, X. Jin, K. Wang, Lighting virtual objects in a single image via coarse scene understanding. Sci. China Inf. Sci. **57**(9), 092105 (2014). https://doi.org/10.1007/s11432-013-4936-0
18. J. Husár, L. Knapčíková, Implementation of augmented reality in smart engineering manufacturing: Literature review. Mobile Netw. Appl. (2023). https://doi.org/10.1007/s11036-023-02121-x
19. E. Kucera, O. Haffner, R. Leskovský, Interactive and virtual/mixed reality applications for mechatronics education developed in unity engine, in 2018 Cybernetics & Informatics (K&I), Lazy pod Makytou, Slovakia (2018), pp. 1–5, https://doi.org/10.1109/CYBERI.2018.8337533
20. V.T. Nguyen, T. Dang, Setting up virtual reality and augmented reality learning environment in unity, in 2017 IEEE International Symposium on Mixed and Augmented Reality (ISMAR-Adjunct), Nantes, France (2017), pp. 315–320, https://doi.org/10.1109/ISMAR-Adjunct.2017.97
21. D.Q. Gbadago, J. Moon, S. Hwang, Mixed Reality-based chemical reactor visualization. IFAC-PapersOnLine **55**(7), 580–585 (2022). https://doi.org/10.1016/j.ifacol.2022.07.506
22. P. Maniam, P. Schnell, L. Dan, R. Portelli, C. Erolin, R. Mountain, T. Wilkinson, Exploration of temporal bone anatomy using mixed reality (HoloLens): Development of a mixed reality anatomy teaching resource prototype. J. Vis. Commun. Med. **43**(1), 17–26 (2020). https://doi.org/10.1080/17453054.2019.1671813
23. A. Maffei, M. Dalle Mura, F.M. Monetti, E. Boffa, Dynamic mixed reality assembly guidance using optical recognition methods. Appl. Sci. **13**, 1760 (2023). https://doi.org/10.3390/app13031760
24. B.A.D. Marques, E.W.G. Clua, C.N. Vasconcelos, Deep spherical harmonics light probe estimator for mixed reality games. Comput. Graph. **76**, 96–106 (2018). https://doi.org/10.1016/j.cag.2018.09.003
25. S. Hrehova, L. Knapcikova, The study of machine learning assisted the design of selected composites properties. Appl. Sci. **12**, 10863 (2022). https://doi.org/10.3390/app122110863
26. J. Mascenik, Implementation of the designed program for calculation and check of chain gears. MM Sci. J. **2019**, 3431–3434 (2019). https://doi.org/10.17973/MMSJ.2019_12_2019008
27. M. Wieczorowski, J. Trojanowska, O. Sokolov, Digital and information technologies in Metrology 4.0, in *Advances in Design, Simulation and Manufacturing VI. DSMIE 2023*, Lecture notes in mechanical engineering, ed. by V. Ivanov, J. Trojanowska, I. Pavlenko, E. Rauch, J. Pitel', (Springer, Cham, 2023). https://doi.org/10.1007/978-3-031-32767-4_8
28. P. Buň, J. Husár, J. Kaščak, Hand tracking in extended reality educational applications, in *Advances in Manufacturing III. MANUFACTURING 2022*, Lecture notes in mechanical engineering, ed. by J. Trojanowska, A. Kujawińska, J. Machado, I. Pavlenko, (Springer, Cham, 2022). https://doi.org/10.1007/978-3-030-99310-8_25

Sustainable Construction: Circular Materials and Their Impacts on Aspects of Construction Processes

Jana Smetanková, Rastislav Ručinský, Lucia Zemanová, and Gregor Lazor

1 Introduction

Population growth and the concomitant increase in demand for services and goods is leading to a consequent depletion of the given stocks of resources and energy worldwide. The reality of resource scarcity is becoming more and more apparent, waste is increasing rapidly, and the population is facing increasing challenges in terms of material security [1].

The circular economy is becoming an increasingly popular topic. In recent years, the European Commission has adopted several legislative proposals aimed at encouraging Member States to move away from a linear model of production and consumption toward a new model aimed at closing the industrial flow of materials. The aim of the circular economy is to recover and preserve a given quality and value of a product or material over as long a cycle as possible. The main objective of circular economics is to minimize overall waste and to use resources efficiently [2].

The circular economy represents the ideal state, that is, an economy without waste, so-called zero waste. The main advantages of a circular economy include environmental protection (reducing overall greenhouse gas emissions), reducing the amount of waste produced, and reducing input costs through the use of recycled

J. Smetanková (✉) · R. Ručinský
Technical University of Kosice, Faculty of Civil Engineering, Expert's Institute in Construction, Kosice, Slovakia
e-mail: jana.smetankova@tuke.sk; rastislav.rucinsky@tuke.sk

L. Zemanová · G. Lazor
Technical University of Kosice, Faculty of Civil Engineering, Institute of Construction Technology, Economics and Management, Kosice, Slovakia
e-mail: lucia.zemanova@tuke.sk; gregor.lazor@student.tuke.sk

L. Knapčíková, D. Peraković (eds.), *8th EAI International Conference on Management of Manufacturing Systems*, EAI/Springer Innovations in Communication and Computing, https://doi.org/10.1007/978-3-031-53161-3_11

materials. The circular economy can create new markets, new opportunities for profit or innovation.

If we compare the linear model of the economy with the circular economy, the circular economy separates the need to extract new materials from economic growth. In practical terms, the circular economy model consists in saving materials by their reuse. These processes lead to a reduction of material orderability or reuse in the product or material cycle. The difference between linear and circular economy is also that the circular economy is a set of principles and tools whose aim is to contribute to the sustainability of the planet by minimizing the extraction and degradation of materials, thus saving energy resources, promoting the regeneration of its input resources, and promoting the use of waste as a primary resource [3].

Improving circular economy practices in the construction industry is key to increasing material reuse and ensuring high quality. Data on waste production in the construction industry show that the situation has been relatively stable at European level in recent years, but there are large differences in per capita production. The high volume and nature of the waste produced is challenging to manage, but at the same time presents an opportunity to implement good waste management practices in line with the principles of circular economy [4].

There is still a lack of emphasis on promoting circularity in the construction industry, hence the need for increased emphasis on creating circular resource flows through reuse and recycling, expanding the phases of use and recycling of materials through refurbishment and remanufacturing, and reducing the use of natural resources and maximizing the efficiency of production processes [4].

A growing number of architects, engineers, and clients are experimenting with circular design. From day one, they have been trying to establish the principle that construction waste should not end up in landfills or as fill under highways, but should be used in construction. So far, however, circular construction has mostly not gone beyond the pilot project stage [4].

The transition to circular construction requires a clear and sustained demand that can drive the market. This applies to both building materials and construction processes. There are two main reasons for this: first, the market is new, and second, the circular principles have not yet been implemented in the rules on procurement procedures, standards, certifications, and the Executive Order on harmful substances [4].

Based on the above, the professional public's perspective on the choice of materials is changing, as the individual actors are not only looking at the basic aspects (cost, deliverability, etc.) but also at the impacts on selected environmental aspects (amount of CO_2 emissions, SO_2, social taxes, recycling rates, etc.).

2 Circular Economic and Construction Industry

The planet's resources are a finite entity. In order to protect the environment, it is necessary to apply sustainable and efficient ways of processing raw materials.

The basic principles of circularity are as follows:

- *Promoting circularity* – the emphasis is on the use of materials that are a renewable resource or can be reused.
- *Supporting the development of recycling technologies on a commercial scale* – the aim is to address the huge problems facing the industry.
- *Promoting the use of technologies and products* – optimizing the processing, use, recovery of chemicals, water, and energy.
- *Promoting transparency and follow of supply chains.*
- *Promoting partnerships* across the value chain to drive systemic change with an emphasis on sustainability.
- *Waste management* – waste generated in the production process can be used as raw material for other processes, thereby reducing the need for new resources, avoiding waste, and increasing resource efficiency [5].

In 2020, the total waste generated in the European Union by all economic activities and households was 2135 million tonnes or 4815 kg per capita. In the EU, construction accounted for 37.5% of the total in 2020. The next positions were occupied by mining and quarrying (23.4%), waste and water services (10.8%), manufacturing (10.6%), and households (9.4%). The remaining 8.2% was accounted for by waste from other economic activities (services – 4.4%, energy – 2.3%). In the EU Member States, waste production in 2020, with the exception of large mineral waste, averaged 6.3 tonnes per capita. A large amount of waste generated in Estonia is caused by the extractive industry or energy production (oil shale). Slovakia is ranked 22nd in the above ranking. The amount of recovered waste (waste recycling and reuse) increased by 33.9% from 870 million tonnes of waste in 2004 to 1165 million tonnes in 2020. On this basis, the share of such recovery in total waste treatment increases from 45.9% in 2004 to 59.1% in 2020. The amount of waste to be disposed of has decreased by 21.5%, while the share of disposal in total waste treatment has decreased from 54.1% (2004) to 40.9% (2020) [6].

Waste management policies aim to reduce the adverse environmental impacts of waste and thereby improve health and resource efficiency in Europe. The Green Deal is the EU's essential tool to support growth in the transition to a modern, resource-efficient and competitive economy.

2.1 Application of Circularity Tools in Construction Industry

With the growing interest in transforming the economy into a more circular version, it is necessary to define the basic tools and techniques. Materials are only one type of resource in our economy. All of its flows are interconnected and interact with each other. It is important to design and develop systems that will fully recover all materials through extremely costly and energy-intensive recycling processes. In addition to materials, attention must also be paid to energy, which is currently a barrier. Also, renewable energy is generated by devices that are made of scarce materials that adversely affect the environment. In view of this, scarce resources must be treated with caution. Ultimately, all energy in a circular economy should be supplied from renewable or otherwise sustainable forms.

If we examine all aspects of the circular economy, we find a number of pre-linkages in the framework of the economic system that need to be sorted out. Based on this, it is possible to define seven basic pillars of the circular economy:

1. *Circulation of materials at their invariably high value* – material complexity is maintained by cascading materials in their most complex form for as long as possible. Material cycles are designed to prime-rank length for human and natural cycles. Rare materials are cycled with priority (shorter intervals). Materials are transported over as small a geographic range as possible, not mixed in ways that preclude separation and recovery (unless they can continue in their mixed form in an endless cycle). Materials are only used when it is unavoidable.
2. *Energy based solely on renewable resources* – materials needed for energy production and storage are used to renew the system. Energy is intelligently stored and cascaded. The density of energy consumption is adapted to the density of local energy availability (elimination of structural energy losses due to transport). Do-overs and conversions between energy types are avoided.
3. Biodiversity is enhanced and supported by human activity – a fundamental pillar of complexity and biodiversity conservation. Habitats are not damaged by human activity. Material and energy losses are tolerated in the interests of biodiversity conservation.
4. *Preserving human and cultural society* – in a circular economy, processes and organizations use appropriate management models and ensure continuous information. Activities that undermine the well-being or existence of human cultures are fully eliminated.
5. *Structural promotion of the well-being and health of humans and other species* – elimination of toxic and hazardous substances (minimization, maintenance of highly controlled cycles, complete elimination). Economic activities do not threaten human health or well-being.
6. *Human activities maximize the creation of social value* – materials and energy are finite, so their use should contribute to the creation of social value. Forms of value go beyond financial value and include aesthetic, emotional, ecological, social, and other values. The choice of resource use maximizes value creation in as many categories as possible and maximizes financial return.
7. Water resources are sourced and cycled sustainably – sufficient water quantity and quality is essential for our economy and our survival. In a circular economy, the value of water is preserved through cycles and unlimited reuse, simultaneously recovering valuable resources. Protecting watersheds is a priority and preventing harmful emissions into aquatic ecosystems is a top priority [7].

The priority is therefore to reuse materials. In a circular economy, materials must go through as few process steps as possible for reuse, thus ensuring a higher quality of the remaining material [8].

We know many types of materials that can be applied to the circular economy from construction and demolition projects. These are mainly gypsum board, plastics, glass, wood, bricks, floor and wall coverings, and insulation and packaging of building materials [9].

Cement, aluminum, steel, and plastic are four of the five commodities that account for up to 55% of global industrial carbon emissions. Their main end-user is the construction industry. To achieve a net zero transition, the supply of and access to circular materials need to be strengthened. Circular material can be defined as material that has already been pre-used that is low or zero carbon, recycled or recyclable [10].

Plasterboard is a material that is quite often used in the construction industry, especially for interior finishing and building partition structures. After separation of plasterboard waste from construction and demolition waste, it can be reused in the production of gypsum from natural or synthetic sources (cement, plaster) or reused in the food and toilet industry [11].

Plastics are versatile materials, but the way they are used is incredibly wasteful. One of the most efficient ways of recycling plastic for subsequent use is chemical recycling. Unlike mechanical recycling, which is inevitably limited by contamination and complications arising from material separation, chemical recycling is a process in which the material is stripped of its original chemical building blocks so that they can be reincorporated into new products. For a long time, chemical recycling often remained on the fringes of the public debate on plastic waste (mostly due to lack of investment and infrastructure), but is nowadays at the center of intense innovation [12].

In the construction industry, plastics can be reused mainly for pipes, insulation, wall and floor cladding, interior fittings, window frames, scaffolding boards, and kerbs [13].

Glass is a pioneer in the field of circulating materials. Glass is a natural product made from purely natural materials (sand, soda, lime, shards) and is therefore infinitely recyclable. Recycled glass can be reused in construction as a substitute for aggregate in road building, for decorative materials, insulation, containers, sports turf, flux in brick and ceramics production, and more [14].

Most types of wood can be recycled. Waste wood in the construction pre-industry can come in a variety of forms from medium density fiberboard to tree branches. Wood and wood waste that comes from the construction industry can be applied to the circular economy as particleboard and fiberboard, bedding production, production of play surfaces and pavements, production of remanufactured products such as fiber composites, production of architectural components, and landscaping [15].

Brick is a building material that is used to make walls, paving, and brick structures. According to the Institution of Civil Engineers' Demolition Protocol (ICE, 2008), bricks have a 10% restoration potential, which can increase to 100% in certain buildings [16]. Waste bricks can be used as a substitute for natural aggregate, a partial replacement for cement or a complete replacement for cement by means of alkaline activation.

2.2 *Impact of the Circular Economy on the Construction Industry*

The need to construct huge numbers of new buildings because of the rapid increase in the planet's population and also the increase in the cost of living is at odds with the need to tackle global warming and the decline in biodiversity. To overcome this twin challenge, it is necessary to move toward a circular economy, for example, by making extensive use of a combination of alternative low-carbon building materials, alternative technologies and practices [17].

The building sector generates approximately 20% of the world's total anthropogenic greenhouse gas emissions [18] and uses a large amount of natural resources. For example, sand and gravel are the most mined group of materials [19] and their subsequent extraction has a major impact on biodiversity [20].

The construction sector is responsible for approximately 50% of the waste generated in the European Union and 25% of solid waste worldwide. The management of this waste has a very negative impact on our environment, among such waste, up to approximately 75% is made up of natural soils and stones [21, 22].

Environmental changes and declining asset availability are forcing major infrastructure pools to seek greater flexibility and efficiency in organizing, improving, and transporting major projects around the world. One of the flexible approaches under consideration is the concept of a circular economy model. The construction and engineering industry is the world's largest purchaser of natural materials. It accounts for half of the world's steel production and burns more than three billion huge quantities of natural substances. Improving resource use can enable the built environment to make more productive use of infrastructure, vehicles, and space. In the built environment, resource owners can lease or share underutilized space, construction and development materials, and equipment. By consuming less space and limiting the amount of time a resource is inactive, fewer assets are expected to provide similar capacity or administration and therefore less waste is generated [23].

The built environment is the construction sector with the largest volume of material flow in the entire global economy. The main significance is that the concept of the circular economy has the potential for fundamental change, particularly in the construction sector projects, namely, civil engineering and infrastructure development. The main principle is to preserve the maximum value of resources, materials, and at the same time the possibility to avoid losses from the system [24].

Based on the above, it is important to follow these five key aspects:

1. Reuse of materials
2. Investigation and analysis of the application of circular economy principles in the construction industry
3. The concept of "material belts" for the tracking and use of resources within the economy
4. Cost implications for buildings
5. Analysis of the structure and values of the European Commission's Circular Economy Package [25]

The circular economy can provide an incentive for achieving sustainable goals or sustainable construction. The construction industry is characterized by strong project-based institutionalized practices as well as market mechanisms that in many aspects do not correspond to the principles of the circular economy [26]. Combining different life cycle design and construction strategies for materials and components could provide a simple way to measure circular economy performance at the design stage of a project and perhaps reduce the complexity of applying circular economy in construction, making it more understandable to different parties in the supply chain. The circular economy can also help facilitate the alignment of a broad agreed definition of circular economy within the built environment that takes into account short- and long-term goals and benefits, which can motivate collaboration between different industry stakeholders in a complex supply chain.

3 Methodology

3.1 Research Aim

The research focuses on the analysis of the application of the basic principles of circular construction in building practice. Based on the theoretical analysis, the main objective of the research was set, specifically the aim was to compare the impacts of using primary and circular materials in the reconstruction of a health center from an economic, environmental, and social point of view.

3.2 Data Collection

The aim of the research was to create a variant solution for the renovation of the health center with an emphasis on taking into account the principles and tools of circular economy. On the basis of the investor's requirements and after studying the project documentation, variant solutions for the implementation of the reconstruction were defined. The data or information on the prices of materials were obtained from quotations of construction companies located near the construction site (radian up to 10 km), for the environmental and social analysis, the data and outputs were the result of work in the Internet tool – OneClick LCA software.

3.3 Research Step and Methodology

The first step of the research was to establish the basic problem and define the main objective of the research. After defining the main objective, the basic evaluation criteria were set, that is, economic, environmental, and social comparisons. Based on the above, the following research steps were implemented:

- Defining the scope of the reconstruction work (defining individual processes and materials)
- Analysis and definition of substitution of standard materials for their circular material variant
- Economic comparison of selected alternative solutions for the renovation of the health center
- Environmental comparison of selected alternative solutions for the renovation of the health center
- Social comparison of selected alternative solutions for the renovation of the health center
- Drawing conclusions

4 Case Study – Application of Circular Construction in the Renovation of a Health Centre in the Spišská Nová Ves

The subject of the assessment, respectively case study was implemented in the framework of the renovation of the health center, which is located in the cadastral territory of Spišská Nová Ves. The building is located in the urbanistic part of the housing estate, where the architecture of prefabricated apartment buildings with flat roofs prevails. Construction modifications within the reconstruction concern changes in the layout, with intervention in the load-bearing and perimeter structures, building an entrance ramp and a staircase platform for immobile patients, replacement of floor layers, ceramic tiles, and wall and ceiling finishes; replacement of medical pre-meters and newly built water supply, sewerage pipes and distribution, external insulation of the building; and replacement of windows and doors.

4.1 Variant Solutions of Reconstruction – Use of Standard and Circulating Materials

Nowadays, sustainability is becoming an important factor in the choice of building materials. Builders and architects are looking for ways to minimize the negative environmental impacts of construction activities while achieving quality and functional results. When selecting circular substitutes to standard material, regional aspects or availability of materials in the Slovak Republic were taken into account. The priority was to find suitable alternatives, so-called circular variants of single-individual solutions, on the basis of the project documentation. Emphasis was placed primarily on the use of recycled materials, or materials that can be considered as a so-called circular variant (i.e., with a higher recycling potential, or in the case of the abovementioned materials, it is possible to recreate several life cycles). Based on the above key, all standard materials have been replaced by their circular variant, for example:

- Grout has been replaced by eco-grout (eco-grout is a mortar made from less energy-intensive materials, such as clay mortar).
- The polyurethane separation film was replaced by a film made of biodegradable material.
- The plastic sill plate was replaced by a sill plate made of composite wood.
- OSB (Isover) was replaced by OSB made of recycled materials (Napkollektor).
- Plastic doors and windows were replaced by wooden doors and windows made of wood with FSC (Forest Stewardship Council) certification.
- Ceramic tiles replaced by tiles made of technical ceramics (recycled ceramics), etc.

4.2 Economic Comparison of Alternative Solutions

Economic analysis or determination of construction costs is one of the key processes across the life cycle of a building. The correct determination of the cost, the price of the building, is an important indicator when deciding on the shape, size, functions, and the actual implementation.

The subject of valuation of construction production are materials, products, semi-finished products, construction objects, operating volumes of construction works, transport performance, and machine performance. In the economic comparison of the alternative solutions, only the material costs were compared with each other. In the economic comparison, the more financially advantageous variant is the reconstruction solution using standard materials. The above costs amount to EUR 85,337.71. The increase would be EUR 76,321.71, which is 52.79% compared to the option using standard materials. Out of the total number of items (69 valued materials), only five items were cheaper in the 2-hour option (representing about 7.25% of the total number of items) (see Table 1). Lower cost of circulating materials was recorded for ecological mortar, acrylic sealant A-1000, partitions made of recycled wood, recycled plastic mesh, and aluminum ladder. A detailed comparison of the above materials with their standard equivalent is shown in Table 2.

4.3 Environmental Comparison of Alternative Solutions

Environmental sustainability is increasingly resonating in society. Sustainability is a feature of construction that can reduce the burden on the climate, and we will implement the construction with the requirements of the Earth's development.

Table 1 Economic comparison – standard and circulating materials

	Costs (EUR)	Costs difference (EUR) – circulating vs. standard materials	Costs – difference (%) circulating vs. standard materials	Total items (construction materials)	Number of items – lower cost (compared to the second option)	% – Number of items lower cost (compared to the second option)	Summary (economic comparison)
Standard materials	85,337.71	+76,321.71	+52.79	69	64	92.75	**WINNER**
Circulating materials	161,659.42			69	5	7.25	*LOSER*

Table 2 Economic comparison – selection of circulating materials with lower costs

Title of material	Unit price of material	Price – difference (EUR) – circulating vs. standard materials	Price – difference (%) – circulating vs. standard materials
Grouting mortar (standard material)	92.42	**−50.27**	**−54.40**
Ecological mortar (circulating materials)	42.15		
Asphalt varnish ALP-PENETRAL SN (standard material)	24.19	**−11.19**	**−46.27**
Acrylic sealant A-1000 (circulating material)	13.00		
Separating partitions (standard material)	4395.52	**−3071.69**	**−69.88**
Partitions made of recycled wood (circulating material)	1313.83		
Grid metal (standard material)	1300.00	**−969.75**	**−74.59**
Recycled plastic mesh (circulating material)	330.25		
Ladder galvanized (standard material)	110.00	**−10**	**−9.09**
Aluminum ladder (circulating material)	100.00		

Modern architecture that is associated with sustainable buildings is becoming more and more popular and sought after across all generations. The aim of the construction company is to produce and use materials with low environmental impact and in turn with a positive effect in their application. The Earth's ecosystems are currently at a critical stage as they are severely damaged and human activities lead to irreversible ecosystem losses. Buildings and construction activities are the largest contributors to global resource use and emissions pollution. In OECD countries, the built environment is responsible for around 25–40% of total energy consumption, 30% of raw materials used, 30–40% of global greenhouse gas emissions and 30–40% of solid waste [27].

On the basis of the above, it was necessary to compare the alternative solutions also from an environmental point of view. As part of this comparison, standard and circulating materials that could be used in the renovation of the health center were compared. The comparison was carried out in the online software tool OneClick LCA. The processing of the environmental analysis was handled in the following steps:

1. Identification of individual material groups and demonstration of information about their properties and life cycle.
2. Inputting the materials data into the LCA software and then calculating the environmental impacts for each phase of the life cycle.
3. Defining and calculating the amount of greenhouse gas emissions, water, and energy consumption for each phase of the life cycle of the material in the LCA software.
4. Comparison of environmental impacts. Analysis of LCA results and comparison of environmental impacts.

In the environmental comparison, the more advantageous option is the reconstruction solution using circulating materials. The amount of carbon footprint is 288.52% lower using circulating materials (−176.00 tons CO_2 less). Out of the total number of items (69 evaluated materials), 65 entries were environmentally preferable with the circular option (representing about 94.20% of the total number of items) (see Table 3). Lower CO_2 pollution was recorded for four items, namely, ceramic prestressed lintel, anchoring element, EPS gradient plates, and separating partitions. A detailed comparison of these materials with their circulating equivalent is shown in Table 4.

4.4 Social Comparison of Alternative Solutions

The social cost of carbon dioxide (SCC) is a social indicator that measures the social cost of emitting CO_2 and other greenhouse gases into the atmosphere. The calculation of the social cost is based on the assumption of the impact of greenhouse gas emissions on climate change and seeks to express the total social cost in euros per ton of CO_2 emissions.

In the case of the renovation of the health center in Spišská Nová Ves, a Life Cycle Assessment (LCA) program was used, which is a method for measuring the environmental impacts of products or services. The SCC calculation in the LCA takes into account health costs, environmental damage, and economic losses due to CO_2 and other greenhouse gas emissions.

Table 3 Environmental comparison – standard and circulating materials

	CO_2 – carbon foot-print (TON)	CO_2 – carbon footprint – difference (TON) – circulating vs. standard materials	CO_2 – carbon footprint – difference (%) – circulating vs. standard materials	Total items (construction materials)	Number of items – lower carbon footprint (compared to the second option)	% – Number of items – lower carbon footprint (compared to the second option)	Summary (environmental comparison)
Standard materials	237.00	**−176.00**	**−288.52**	69	4	5.80	*LOSER*
Circulating materials	61.00			69	65	94.20	**WINNER**

Table 4 Environmental comparison – selection of standard materials with lower carbon footprint – emissions CO_2

Title of material	Carbon footprint	Carbon footprint – difference (TON) – circulating vs. standard materials	Carbon footprint – difference (%) – circulating vs. standard materials
Ceramic prestressed lintel (standard material)	34.00	*+42.00*	*+55.26*
Glass pre-stressed lintel (circulating materials)	53.00		
Anchoring element (standard material)	0.48	*+47.52*	*+99.00*
Anchoring element based on recycled materials (circulating material)	48.00		
EPS gradient plates (standard material)	72.00	*+178.0*	*+71.20*
Falling boards made of recycled material (circulating material)	250.00		
Separating partitions (standard material)	790.00	*+160.00*	*+16.84*
Partitions made of recycled wood (circulating material)	950.00		

In the social comparison, the more favorable option is the reconstruction solution using circulating materials. The social cost of using circulating materials is EUR 3030.00, which is almost 74% less than using standard materials (see Table 5). By using a more sustainable option, it is possible to reduce the societal cost of carbon dioxide, which has a significant impact on environmental protection. CO_2 emissions degrade air quality and can lead to an increase in respiratory diseases and deaths. At the same time, the analysis also generated a "Cradle to Grave" certificate

Table 5 Social comparison – standard and circulating materials

Social costs (EUR)	Social costs – difference (EUR) – circulating vs. standard materials	Social costs – difference (%) – circulating vs. standard materials	Summary (social comparison)
11,530.00	**−8500.00**	**−73.72**	*LOSER*
3030.00			**WINNER**

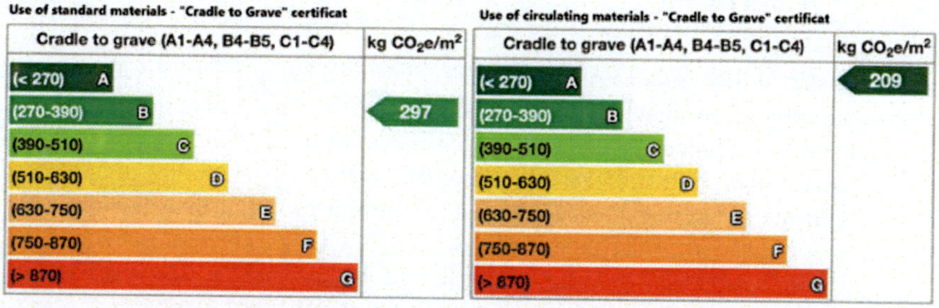

Fig. 1 Comparison of "Cradle to grave" certificates – standard versus circulating materials

to assess the environmental impact of the materials, where reconstruction using circular materials scored better (for detailed results – comparison, see Fig. 1).

4.5 *Summary Comparison and Discussion of Alternative Solutions*

The aim of the above analysis was to assess the impacts of using circular materials in the reconstruction of the health center from an economic, environmental, and social point of view. The above criteria were selected based on an analysis of the basic aspects and pillars to promote sustainable construction.

Based on the analyses carried out, it has been shown that the use of recycled materials can reduce environmental and social burdens. These aspects are key tools to promote sustainability and circularity of materials within the life cycle of buildings. The only aspect where the circular option was a disadvantageous option was the comparison in terms of procurement costs. The final choice or implementation method (use of standard or circular materials) is at the discretion of the investor, based on their chosen criteria. From a sustainability point of view, the use of recycled materials is the optimal choice, but nowadays, there is an increasing emphasis on the economic aspect, where the optimal option is to carry out the reconstruction using standard materials (see Table 6).

Table 6 Summary comparison – use of standard and circulating materials

	Economic comparison	Environmental comparison	Social comparison
Standard materials	**WINNER**	*LOSER*	*LOSER*
Circulating materials	*LOSER*	**WINNER**	**WINNER**

5 Conclusion

Sustainability in construction has become increasingly prominent in recent years, bringing a conscious approach to construction and a positive impact on the environment. The choice of materials in construction can be influenced by various criteria, such as the price–performance ratio. In terms of sustainability of construction and with the increasing pressure to reduce carbon footprint or waste production, it is important to consider other criteria such as CO_2 production, recycling, and reuse options. Based on the above, the use of bi-materials for construction is a more favorable option, but it is important to point out that this option is less favorable from an economic point of view, which may weigh in the investor's decision-making. Creating sustainable buildings forms the basis of a sustainable vision, but this must be linked to a sustainable way of life and its users. These changes can improve the environment and promote the social well-being of the population. Based on the above, it is important to increasingly implement recycled materials in various aspects of the construction industry. It is proven that recycled materials help to reduce the adverse impacts of industry on the Earth's climate. The aim of future research will be to raise awareness of the recycling and reuse potential of selected building materials, particularly through the use of a materials catalogue. The catalogue will exclusively contain materials that can be recycled and will define the possibilities of reuse and will include selected economic (cost), environmental (amount of CO_2 production), and social aspects that will help to select the right material based on the user's preferences.

Acknowledgments This paper presents partial research results of projects APVV-17-0549 "Research of knowledge-based and virtual technologies for intelligent designing and realization of building projects with emphasis on economic efficiency and sustainability" and APVV-22-0576 "Research of digital technologies and building information modeling tools for designing and evaluating the sustainability parameters of building structures in the context of decarbonization and circular construction." This article was created thanks to support under the Operational Program Integrated Infrastructure for the project: National infrastructure for supporting technology transfer in Slovakia II – NITT SK II, co-financed by the European Regional Development Fund.

References

1. C.D.F. Rogers et al., Resource scarcity and resource security – A suppressed civil engineering challenge. Proc. Inst. Civ. Eng. Waste Resour. Manag. **170**(2), 49–51 (2017) Retrieved from https://doi.org/10.1680/jwarm.17.00008. Last accessed 22 Oct 2022

2. T. Guštafikova, A. Kosturikova, Z. Lieskovska, *Circulating Economy – The Future of Slovakia's Development (Obehové hospodárstvo – budúcnosť rozvoja Slovenska)* (Ministry of the Environment of the Slovak Republic, Bratislava, 2019) 104 s. ISBN 978-80-8213-001-3. Retrieved from https://www.enviroportal.sk/uploads/report/9202.pdf. Last accessed 15 May 2023
3. P.G. Palafox-Alcantar, D.V.L. Hunt, C.D.F. Rogers, The complementary use of game theory for the circular economy: A review of waste management decision making methods in civil engineering. Waste Manag. **102**, 598–612 (2020) Retrieved from https://doi.org/10.1016/j.wasman.2019.11.014. Last accessed 1 Oct 2022
4. DTU.dk, Sustainability: Circular construction (2023). Retrieved from https://baeredygtighed.dtu.dk/en/teknologi/cirkulaert-byggeri. Last accessed 19 Aug 2023
5. Lenzing, Circular economy: Economy with a future (2022). Retrieved from https://www.lenzing.com/sustainability/sustainability-management/circular-economy?gclid=CjwKCAjw-IW-BhBTEiwA2exyO7QNBD20xSTTRzFEDvmzEaD_E01rFU7zbjaSrvPJi3Ia3CFgI0_gOhoCveAQAvD_BwE. Last accessed 10 June 2023
6. Eurostat, Waste statistic (2023). Retrieved from https://ec.europa.eu/eurostat/statistics-explained/index.php?title=Waste_statistics#Total_waste_generation. Last accessed 5 June 2023
7. E. Gladek, *The Seven Pilliars of the Circular Economy* (Metabolic, Amsterdam, 2017) Retrieved from https://www.metabolic.nl/news/the-seven-pillars-of-the-circular-economy/?gad=1&gclid=CjwKCAjw4ZWkBhA4EiwAVJXwqR11zlMLkKw4ohzWnPZvNqw0vhxq7Njbtj51kJJuItR1yjYSVlthJBoCx5kQAvD_BwE. Last accessed 5 June 2023
8. Kenniskaarten, How do materials circulate in a circular economy? (2022). Retrieved from https://kenniskaarten.hetgroenebrein.nl/en/knowledge-map-circulareconomy/how-materials-circulate/. Last accessed 19 Apr 2023
9. NiBusiness info, Waste contractor's guide to recycling construction waste, Construction materials that can be recycled (2022). Retrieved from https://www.nibusinessinfo.co.uk/content/construction-materials-can-be-recycled. Last accessed 10 May 2023
10. World Economic Forum, Here's how to create a circular system for the new builtenvironment (2022). Retrieved from https://www.weforum.org/agenda/2022/11/heres-howcircular-system-building-materials/. Last accessed 20 May 2023
11. NiBusiness info, Recycling options for types of waste materials (2022). Retrieved from https://www.nibusinessinfo.co.uk/content/recycling-plasterboard. Last accessed 28 May 2023
12. The SustanAbility Institute by ERM, Creating a circular economy for plastics (2022). Retrieved from https://www.sustainability.com/thinking/creating-a-circular-economy-for-plastics/. Last accessed 26 Oct 2022
13. NiBusiness info, Recycling construction materials: Recycling plastic from construction projects (2022). Retrieved from https://www.nibusinessinfo.co.uk/content/recycling-plastic-construction-projects. Last accessed 3 Nov 2022
14. NiBusiness info, Recycling construction materials: Recycling glass from construction projects (2022). Retrieved from https://www.nibusinessinfo.co.uk/content/recycling-glass-construction-projects. Last accessed 5 May 2023
15. NiBusiness info, Recycling construction materials: Recycling wood from construction projects (2022). Retrieved from https://www.nibusinessinfo.co.uk/content/recycling-wood-construction-projects. Last accessed 3 Nov 2022
16. J. Fořt, R. Černý, Transition to circular economy in the construction industry: Environmental aspects of waste brick recycling scenarios. Waste Manag. **118**, 510–520 (2020) Retrieved from https://www.sciencedirect.com/science/article/pii/S0956053X20305146. Last accessed 15 Feb 2022
17. J.C. Morel et al., Earth as construction material in the circular economy context: Practitioner perspectives on barriers to overcome. Philos. Trans. R. Soc. B **376**(1834), 1–9 (2021) Retrieved from https://doi.org/10.1098/rstb.2020.0182. Last accessed 3 Oct 2022

18. O. Edenhofer et al., 2015 Summary for policymakers, in *Climate Change 2014: Mitigation of Climate Change*, (Cambridge University Press, Cambridge, 2014) Retrieved from https://archive.ipcc.ch/pdf/assessmentreport/ar5/wg3/ipcc_wg3_ar5_summary-for-policymakers.pdf. Last accessed 5 Oct 2022
19. M. Bendixen et al., Time is running out for sand. Nature **571**, 29–31 (2019) Retrieved from https://doi.org/10.1038/d41586-019-02042-4. Last accessed 30 Sept 2022
20. E. Park et al., Dramatic decrease of flood frequency in the Mekong Delta due to river-bed mining and dyke construction. Sci. Total Environ. **723**, 138066 (2020) Retrieved from https://doi.org/10.1016/j.scitotenv.2020.138066. Last accessed 30 Sept 2022
21. G.L.F. Benachio, M. Do Carmo Duarte Freitas, S.F. Tavares, Circular economy in the construction industry: A systematic literature review. J. Clean. Prod. **260**, 121046 (2020) Retrieved from https://doi.org/10.1016/j.jclepro.2020.121046. Last accessed 30 Sept 2022
22. C. Eras et al., Improving the environmental performance of an earthwork project using cleaner production strategies. J. Clean. Prod. **47**, 368–376 (2013)
23. CivilBites.com, Is it any environment benefit of the circular economy for civil engineering? (2022). Retrieved from https://www.civilsbites.com/2021/11/12/is-it-anyenvironment-benefit-of-the-circular-economy-for-civil-engineering/. Last accessed 3 Nov 2022
24. S. Bringezu, Possible target corridor for sustainable use of global material resources. Resources **4**, 25–54 (2015) ISSN 2079-9276. Retrieved from https://doi.org/10.3390/resources4010025. Last accessed 24 Oct 2022
25. J. Hill, The circular economy: From waste to resource stewardship, part I. Waste Resour. Manag. **168**(1), 3–13 (2015) Retrieved from https://doi.org/10.1680/warm.14.00003. Last accessed 13 Oct 2022
26. Ellen Macarthur Foundation, Growth within: A circular economy vision for a competitive Europe (2015) Retrieved from https://www.mckinsey.com/capabilities/sustainability/our-insights/growth-within-a-circular-economy-vision-for-a-competitive-europe. Last accessed 22 Dec 2022
27. Saint Gobain: ISOVER, Sustainability (Trvalá udržateľnosť) (2023). Retrieved from https://www.isover.sk/trvala-udrzatelnost. Last accessed 10 June 2023

Project Management and Indirect Operating Costs by Building Information Modeling in Construction Projects in Slovakia, Slovenia, and Croatia

Tomáš Mandičák, Annamária Behúnová, Marcel Behún, and Peter Mésároš

1 Introduction

In addition to the solution of technical and technological issues, the issue of construction project management also represents the solution to the economic agenda associated with the project [1]. Economic activity, which should be based on a precise determination of quantitative indicators, can be helpful in the effective management of construction projects and in achieving positive performance indicators [2]. Defining key performance indicators also in the management of construction projects is important from the point of view of clearly defined goals, methods, and strategies for their achievement and the use of selected technologies [1]. Selected key performance indicators must be monitored and evaluated [3]. Based on these indicators, it is possible to manage operationally and make decisions to achieve the set goals [4]. This represents one of the primary management tools for efficiently using available resources [5]. Project costs can also be included among these key indicators based on several sources. It is necessary to define and monitored key performance indicators associated with the management of construction projects [6].

Operational or transactional costs represent an essential part of cost indicators. These indicators also represent a proportional part of the costs that must be managed for efficiency [7, 8]. Building information modeling (BIM) technologies currently

T. Mandičák · P. Mésároš
Technical University of Košice, Kosice, Slovak Republic
e-mail: tomas.mandicak@tuke.sk; peter.mesaros@tuke.sk

A. Behúnová (✉) · M. Behún
Faculty of Mining, Ecology, Process Control and Geotechnologies, Institute of Earth Resources, Kosice, Slovak Republic
e-mail: annamaria.behunova@tuke.sk; marcel.behun@tuke.sk

L. Knapčíková, D. Peraković (eds.), *8th EAI International Conference on Management of Manufacturing Systems*, EAI/Springer Innovations in Communication and Computing, https://doi.org/10.1007/978-3-031-53161-3_12

bring great possibilities for streamlining work and achieving better results in the construction industry [9]. This should also impact key performance indicators, which are the bearer of basic information about the project and the possibility of evaluating the project's success. Implementing BIM technology in the management of construction projects brings several challenges, which in some cases represent barriers to a more effective use of this tool [10]. These are primarily technical, psychological, personnel, and economic factors. Economic factors can represent a strong argument for the decision on their more frequent use [11].

On the one hand, there are implementation and procurement costs for selected technologies. On the other hand, the operating costs are the subject of a complex evaluation of the implementation of new technologies of use [12]. Another factor in economic evaluation and the benefits of implementing BIM technology can be the impact on key performance indicators and the search for relationships between them. To a large extent, these are also cost indicators for project management itself, or the so-called transactional or operational indirect costs, i.e., costs including sales, communication and administrative costs [13]. Implementing information and communication technologies (including BIM technology) should also bring this perspective on the effectiveness of the investment [14]. These solutions have several benefits and work, and the use of data and information is one of them. Management and the use of information for project management's needs are important aspects of achieving the set goals in economic terms. Therefore, the fundamental research question is to what extent new technologies, such as the use of BIM in the management of construction projects, bring improvement and results in the economic field, specifically for selected cost items. This issue is a topic on several construction markets. The implementation of new technologies should always be analyzed over time from data from practice, how it helps improve and achieve the set goals and whether it is actually beneficial. In the same way, the questions of when and how significant the benefit is also in the form of an economic reconciliation should be answered.

2 Theoretical Basis

2.1 BIM Technology and the Value of Information in the Management of Construction Projects

Building information modeling (BIM) represents an information model of the entire life cycle, which expresses technical parameters through physical and functional characteristics, which ensures information sharing and decision support from the design phase to the operation and maintenance phases and the end of the life cycle [15]. For information management in construction, BIM represents a significant milestone [16].

BIM represents an approach that ensures the essential functions of information systems [2, 17, 18]: the collection, processing, analysis, and subsequent deployment of complex information during each phase of the construction life cycle [19]. According to studies, the advantage of using BIM is the connection of parameters and objects, which advances the work with information which aims to improve productivity in the construction industry [20]. According to Gao and the collective of authors and other studies [21–24], BIM represents a tool that combines information about construction projects, project management, and last but not least, project performance analysis. Digitization in the construction industry has highlighted collecting and combining data from multiple sources [25]. The construction industry accepts BIM technology for the needs of digital project management and planning throughout the entire life of the project [26].

In some cases, BIM is used to design and implement buildings and operation and ensures these multiple positive effects on the economy, society, and the environment [27]. In addition to several advantages that BIM technology brings, positives can also be classified in the field of quality management, as this tool also presents opportunities to reduce and manage risks using various international guidelines [26]. BIM is perceived as a means to achieve proper work management [28]. Very interesting are the statistics of the increase in the implementation of BIM technology among selected regions and groupings of the world in the last few years, since 2010 (Fig. 1) [29].

Among the main advantages of using BIM is the provision of appropriate communication and exchange of information. The mentioned study claims BIM reduces disagreements and confrontations [30] due to its added informational value. What is important from the point of view of data presentation, BIM technology brings the possibility of presenting data to the project and the building visually, which can be more attractive to the user [31]. However, as a result, the perception and understanding of data can appear more logical and easier to understand. Digitization in construction and the use of BIM enables the modeling of building

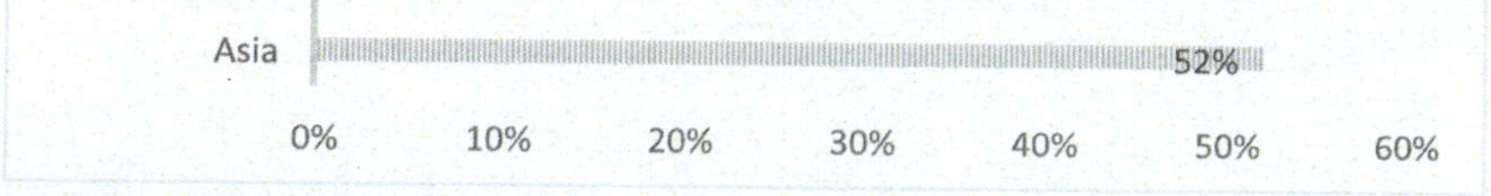

Fig. 1 Increase in adoption and use of BIM technology by individual world regions and geographic groupings

information, which is not considered just a virtual model or software. This is also the process of creating semantically rich information used in the construction industry – an effective platform for information management and technology integration [32, 33].

2.2 Key Performance Indicators, Efficiency Assessment, and Indirect Operating Costs

Every business entity should measure its performance and manage itself on the basis of objective data [3, 17]. Key performance indicators (KPIs) are indicators that can be used to measure the performance of companies or business processes [2]. Likewise, when managing construction projects, it is obvious that, just like a business, it must be managed on the basis of objective data and indicators. Several sources list selected KPIs that define the performance of either companies or projects. For projects, these indicators can be classified into three basic groups: economic, social, and environmental KPIs (Fig. 2).

Performance can be defined as the amount of efficiency and effectiveness in all projects [34]. Key performance indicators (KPIs) are a measure for evaluating project performance. Dynamic project performance measurement refers to the measurement of project performance while the project is ongoing [35, 36]. Several sources focus on cost-effectiveness indicators [35]. Within the definition of cost indicators, the research focuses on operational indirect costs, i.e., costs including sales, communication, and administrative costs.

Economic KPIs	*Social KPIs*	*Environment KPIs*
Client satisfaction – product	*Employee satisfaction*	*Environmental impact*
Client satisfaction – service	*Staff turnover*	*Energy use – product*
Defects	*Sickness absence*	*Energy use – process*
Predictability – Cost	*Safety*	*Water use – product*
Predictability – Time	*Working hours*	*Water use – process*
Safety	*Qualifications & skills*	*Waste removed from site*
Productivity	*Equality and diversity*	*Commercial vehicle movements*
Profitability	*Training*	*Impact on biodiversity*
Construction Cost	*Pay*	*Area of habitat created / retained*
Construction Time	*Investors in people*	*Whole of life performance – product*

Fig. 2 Key performance indicators groups in projects

3 Methodology

3.1 Research Aim and Problem Statement

The research aims to analyze the use of BIM technology in the management of construction projects in the three monitored countries, Slovakia, Croatia, and Slovenia. At the same time, selected performance indicators are monitored in these projects, namely, cost indicators for project management itself or the so-called transactional or operational indirect costs, i.e., costs including sales, communication, and administrative costs. These often represent an equally important cost component that also affects the project's overall cost. These indicators are also compared over time, since this project is ongoing and obtaining data in individual periods was possible, from the beginning of 2015 until 2023.

The research sample included respondents from three countries active in the construction market. The construction market in these countries is relatively small, but it has an essential presence in each country. In Slovakia, the construction industry has a long-term share of approximately 7.5% of the country's GDP. In Croatia, it is less, but even so, it is also an important sector, especially from the point of view of building infrastructure and transport constructions of various natures. The Slovenian construction market is the smallest, but the importance of some projects is relatively high. It is important to note that despite the smaller number of projects, some have similar characteristics to those in Croatia (mainly infrastructure projects). Within the definition of cost indicators, the research focuses on operational indirect costs, i.e., sales, communication, and administrative costs.

3.2 Data Collection and Data Processing

Data collection was ensured through a questionnaire. The questionnaire consisted of essential questions for the characteristics of the research sample, questions regarding the use of BIM technology and questions focused on KPIs, especially cost items. Respondents answered questions based on a Likert scale, where level 1 represented the lowest level of use and level 5, the highest level of use. It was the same with the options for finding the level of KPIs of projects and cost items.

The questionnaire was addressed along with the briefing to the project managers, who were instructed to fill in and understand the questions. It was also strongly recommended to consult with the financial director or the economic department, which has this information as part of the accounting and management of the economic agenda. The research subject was construction projects implemented in the construction markets of three countries. First of all, it was an analysis of the market of these countries, which are similar in size. Slovakia has a predominance of residential buildings and transport structures. In countries like Croatia and Slovenia, transport and infrastructure projects prevailed. More closely, the economic

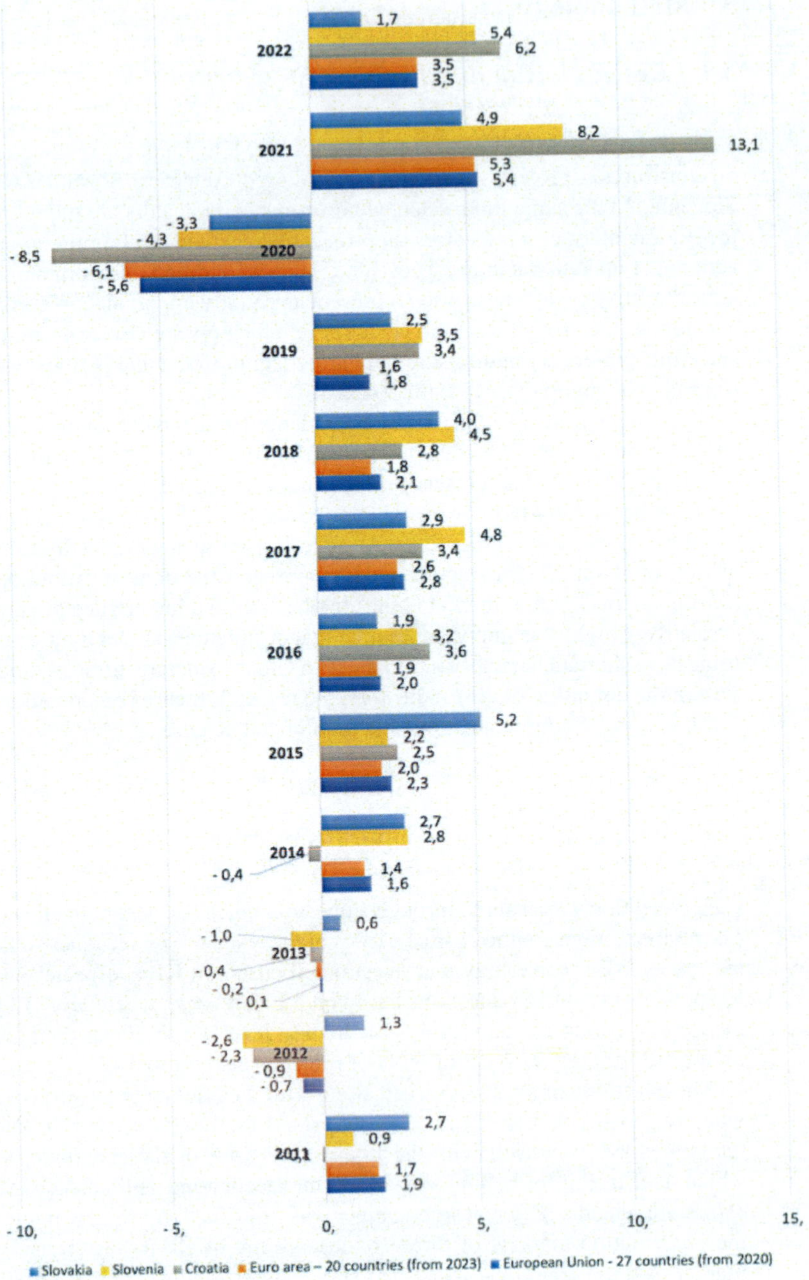

Fig. 3 Comparison a description GDP of research market countries

Table 1 Capital stock-based productivity indicators at industry level (Unit of measure – Index, 2015 = 100)

TIME	2015	2016	2017	2018	2019	2020	2021
Croatia	100.0	101.391	99.031	96.597	95.215	98.729	99.110
Slovenia	100.0	97.416	94.410	91.666	89.859	90.271	89.508
Slovakia	100.0	99.648	99.156	99.025	100.688	103.249	106.337

characteristics of the market and target respondents refer to the following image, which describes Capital stock-based productivity indicators at industry level (Unit of measure – Index, 2015 = 100) (Fig. 3).

The graph shows that this has the highest real GDP growth in the last 2 years. Slovakia and Slovenia are very similar in this indicator. The year 2020 brought not only a drop in production associated with the COVID-19 pandemic but also a drop in real GDP, where the expenses of all components of GDP were dampened due to the impact of the pandemic. Even the decline in the countries in which the research was carried out was more pronounced than the average of the EU countries and the eurozone. Regarding construction production and productivity, the indicators also speak for themselves – the index for the industry, which can be seen in Table 1. The worst results according to this indicator were recorded in Slovakia. On the contrary, the best values were achieved in Slovakia. These indicators indicate the development and situation of power indicators, which also affect the market. From the point of view of the established conditions for the research, it is important to say that these markets and their behavior can be compared, since the indicators they have achieved are similar.

The research sample represents the results of projects obtained from three countries, such as Slovakia, Slovenia, and Croatia. These projects were implemented in different parts of the country and reflect diverse projects implemented in different regions (Fig. 4).

4 Results and Discussion

The research results provide insight into the use of building information modeling and the level of selected cost indicators as part of key performance indicators. These results can largely be related to the use of the given technology and high added information value. Digitization in the construction industry brings many benefits. This is a scientific question whether the impact on key performance indicators is also among those benefits.

Figure 5 provides an overview of the average value of individual cost items and the use of BIM technology. The impact of these technologies was determined at the indirect operating cost at the level of 3.91 and this is a significant indicator and

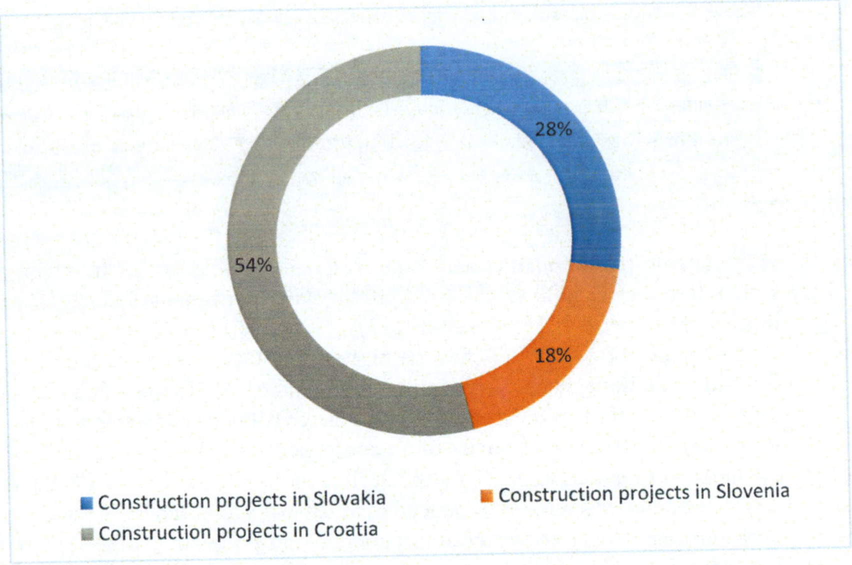

Fig. 4 Research sample by country of construction projects

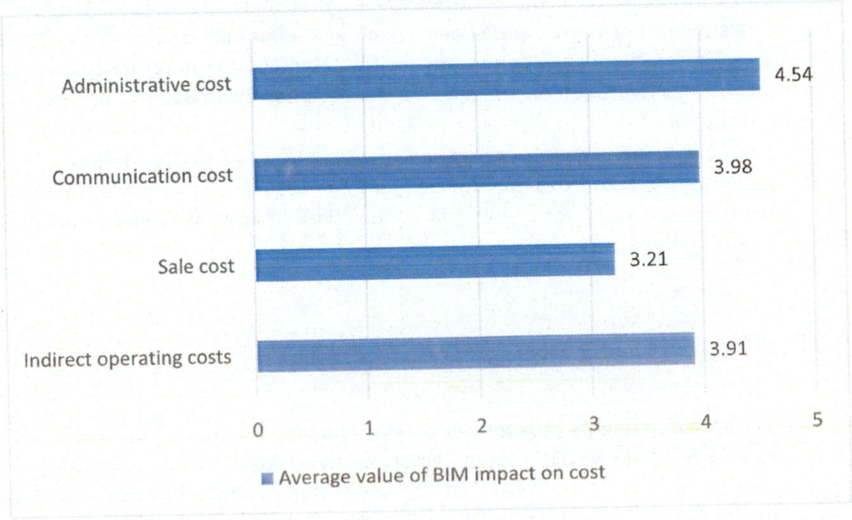

Fig. 5 Impact level of BIM technology on cost issues

value. Administrative cost reached the level of 4.54, sale cost reached the level of 3.21, and communication cost reached the level of 3.98.

An interesting and important comparison of these indicators is shown in Fig. 6. As can be seen in this figure, higher values were recorded for administrative costs

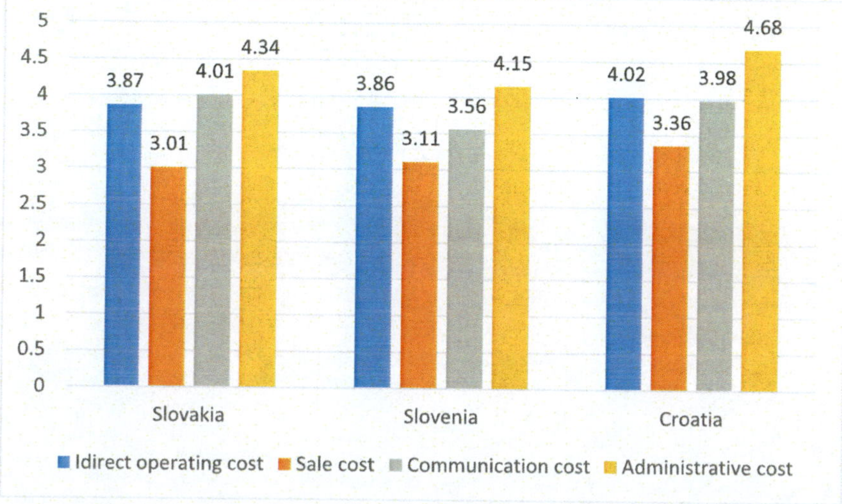

Fig. 6 Cost-reducing level in construction project BIM technology use

in all countries where the research was carried out. In the case of Slovakia, it was at the level of 4.34; in the case of Slovenia, it was at the level of 4.15; and in the case of Croatia, it was at the level of 4.68.

Another important item is the cost of communication, or the impact on them. It also recorded high values at the level of 4.01 for the project in Slovakia, 3.56 in Slovenia, and 3.98 in Croatia. On the contrary, the lowest impact was recorded in the cost of the hall.

Administrative costs contribute to the largest extent to total indirect operating cost. And for most projects, more than 50%. Communication costs are also significant. Sales costs account for more than 10% in most projects (more detailed information in Fig. 7).

When it comes to the use of building information modeling in construction projects, every country has reserves and a lot of room for improvement. Digitization in the construction industry is a prerequisite for improving management as well. The use of information is of great benefit for correct decisions. These informed decisions generally result in effective management. BIM technology has potential, and it is necessary to improve its use in every country. Even in this research, a trend toward this statement was demonstrated.

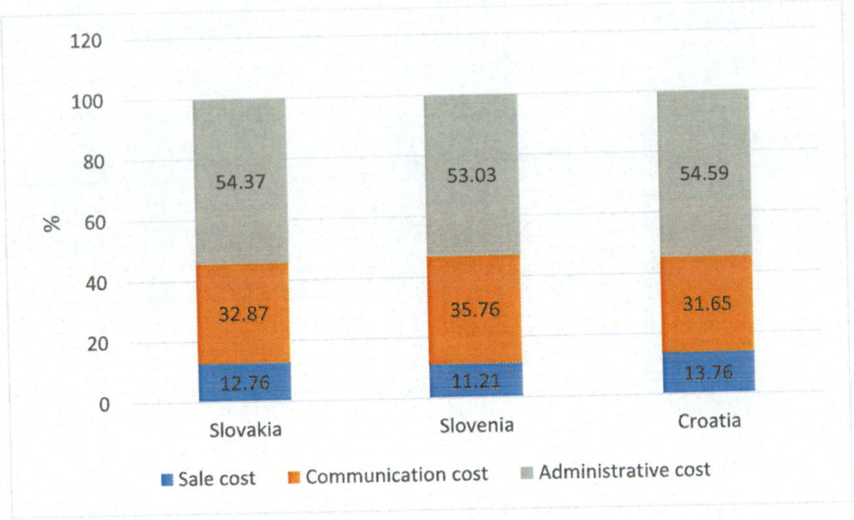

Fig. 7 Share of cost issues on total indirect operating cost (construction project by countries)

5 Conclusion

Research focused on the use of BIM technology and the added value of information in the management of construction projects in Slovakia, Slovenia, and Croatia. At the same time, selected performance indicators were monitored, primarily overhead items and total indirect operating costs. These results provide insight and point to a trend. In general, the research produced several findings. First, the rate of use of BIM technology is similar in these countries, and the investor has the opportunity to influence it to a large extent from the point of view of the participants in the construction project. On the other hand, the potential in the digital age is huge and not fully utilized at all. In other words, there is a relatively large scope for the use of this technology and the use of such important information in the management of construction projects.

The impact on selected cost items as key performance indicators was also analyzed and linked to the level of these results. This means that these indicators were met in these projects. The research pointed to the positive impact of the use of BIM technology in the management of construction projects also in the cost area. Projects where BIM technology was used to a greater extent achieved better results. Above all, this was noted for administrative costs.

The research has great potential for expansion and is not limited to the area of Central and Southeastern Europe. Another research goal is to find out these possibilities and the impact of research in the countries of Western Europe and to map the situation in the entire EU. Another milestone and opportunity is to use the collected data and develop a model of the impact of key performance indicators in construction projects. This can again represent important results not only from the research point of view but also important and valuable for practice. These are further goals for research for the future.

Acknowledgments This paper presents a partial research result of project the Slovak Research and Development Agency under contract no. APVV-17-0549, "Research of knowledge-based and virtual technologies for intelligent designing and realization of building projects with emphasis on economic efficiency and sustainability." This paper also presents a partial research result of project the Slovak Research and Development Agency under contract no. APVV 22-0576 "Research of digital technologies and building information modeling tools for designing and evaluating the sustainability parameters of building structures in the context of decarbonization and circular construction."

References

1. I. Marović, T. Hanak, S. Plaum, Performance management in civil engineering: A systematic literature review. Electron. J. Fac. Civ. Eng. Osijek **13**(24), 47–58 (2022)
2. B. Johansson, F. Sudzina, P. Pucihar, Alignment of business and information strategies and its impact on business performance. J. Bus. Econ. Manag. **15**(5), 886–898 (2014)
3. B. Kršák, Measuring information technology's impact on company performance. Metalurgija **43**(3), 261 (2004)
4. D. Catala-Pérez, M. Rask, M. de-Miguel-Molina, The Demola model as a public policy tool boosting collaboration in innovation: A comparative study between Finland and Spain. Technol. Soc. **63**, 1–17 (2020)
5. B. Dasović, M. Galić, U.A. Klanšek, Survey on integration of optimization and project management tools for sustainable construction scheduling. Sustain. For. **12**, 3405 (2020)
6. S. Kravanja, U. Klanšek, T. Žula, Mass, direct cost and energy life-cycle cost optimization of steel-concrete composite floor structures. Appl. Sci. **11**, 10316 (2021)
7. E. Abdel-Gali, A.H. Ibrahim, A. Alborkan, Assessment of transaction costs for construction projects. Int. J. Constr. Manag. **22**(9), 1618–1631 (2022)
8. L. Downey, G. Scott, Y. Perez, What are transaction costs? Definition, how they work, and example. Investopedia (2023), https://www.investopedia.com/terms/t/transactioncosts.asp
9. Z. Pučko, D. Maučec, N. Šuman, Energy and cost analysis of building envelope components using BIM: A systematic approach. Energies **13**, 2643 (2020). https://doi.org/10.3390/en13102643
10. H. Begić, M. Galić, A systematic review of Construction 4.0 in the context of the BIM 4.0 premise. Buildings **11**, 337 (2021). https://doi.org/10.3390/buildings11080337
11. H. Begić, M. Galić, Z. Dolaček-Alduk, Digitalization and automation in construction project's life-cycle: A review. J. Inf. Technol. Construct. **27**, 441–460 (2022)
12. L. Hermanová, T. Hanák, An empirical analysis of overhead cost management in the Czech construction industry. Tech. J. **11**, 216–2020 (2017)
13. BDC, Operating expenses (selling, general and adminitrsative expenses) (2023), https://www.bdc.ca/en/articles-tools/entrepreneur-toolkit/templates-business-guides/glossary/operating-sg-and-a-expenses
14. O. Olugboyega, Differential relationships in the BIM implementation process in a developing country: The role of essential BIM implementation strategies. Eng. Constr. Archit. Manag. (2023). https://doi.org/10.1108/ECAM-10-2022-0999
15. W. Li, S. Li, Z. Lin, Q. Li, Information modeling if mine working based on BIM technology. Tunn. Undergr. Space Technol. **115**, 103978 (2021)

16. C.M. Eastman et al., Exchange model and exchange object concepts for implementation of national BIM standards. J. Comput. Civ. Eng. **24**(1), 25–34 (2010)
17. G. Gimpel, F. Sudzina, K. Petrovčiková, Mobile ICT use in early adopter vs. late majority countries. Int. J. Mob. Commun. **14**(6), 610–631 (2016) ISSN 1470-949X
18. F. Sudzina, P. Kmec, The technological paradox and evaluation of the benefits of informatization. Econ. J. **54**(3), 281–293 (2006)
19. K.I. Gartoumi, S. Zaki, M. Aboussaleh, Building information modelling (BIM) interoperability for architecture and engineering (AE) of the structural project: A case study. Mater. Today Proc. (2023). https://doi.org/10.1016/j.matpr.2023.05.408
20. K. Idrissi Gartoumi, M. Aboussaleh, S. Zaki, The workability and usefulness of building information modelling based design for building performance. Mater. Today Proc. **58**, 1174–1180 (2022)
21. H. Gao, C. Koch, Y. Wu, Building information modelling based building energy modelling: A review. Appl. Energy **238**, 320–343 (2019). https://doi.org/10.1016/j.apenergy.2019.01.032
22. K. Guo, Q. Li, L. Zhang, X. Wu, BIM-based green building evaluation and optimization: A case study. J. Clean. Prod. **320**, 128824 (2021). https://doi.org/10.1016/j.jclepro.2021.128824
23. M.F. Muller, F. Esmanioto, N. Huber, E.R. Loures, A systematic literature review of interoperability in the green Building Information Modeling lifecycle. J. Clean. Prod. **223**, 397–412 (2019). https://doi.org/10.1016/j.jclepro.2019.03.114
24. M. Rahmani, S. Zarrinmehr, M. Bergin, W. Yan, BPOpt: A framework for BIM-based performance optimization. Energ. Buildings **108**, 401–412 (2015). https://doi.org/10.1016/j.enbuild.2015.09.011
25. X. Huang, Y. Kiu, L. Huang, E. Onstein, C. Merschbrock, BIM and IoT data fusion: The data process model perspective. Autom. Constr. **149**, 1–18 (2023)
26. X. Pan, A.M. Khan, S.M. Eldin, F. Aslam, S.K. Ur Rehman, M. Jameel, BIM adoption in sustainability, energy modelling and implementing using ISO 19650: A review. Ain Shams Eng. J. **15**, 102252 (2024)
27. M.V. Shoubi et al., Reducing the operational energy demand in buildings using building information modeling tools and sustainability approaches. Ain Shams Eng. J. **6**(1), 41–55 (2015)
28. Y.Y. Al-Ashmori et al., BIM benefits and its influence on the BIM implementation in Malaysia. Ain Shams Eng. J. **11**(4), 1013–1019 (2020)
29. O. Adeniyi, N. Thurairajah, F. Leo-Olagbaye, Rethinking digital construction: A study of BIM uptake capability in BIM infant construction industries. Constr. Innov. (2022). https://doi.org/10.1108/CI-09-2021-0161
30. S. Mostafa et al., Exploring the status, benefits, barriers and opportunities of using BIM for advancing prefabrication practice. Int. J. Constr. Manag. **20**(2), 146–156 (2020)
31. J. Li et al., A project-based quantification of BIM benefits. Int. J. Adv. Robot. Syst. **11**(8), 123 (2014)
32. S. Honghong, Y. Gang, L. Haijang, Z. Tian, J. Annan, Digital twin enhanced BIM to shape full life cycle digital transformation for bridge engineering. Autom. Constr. **147**, 104736 (2023)
33. A. Khudhair, H. Li, G. Ren, S. Liu, Towards future BIM technology innovations: A bibliometric analysis of the literature. Appl. Sci. **11**, 1232 (2021)
34. N. Nassar, S. Abou Rizk, Practical application for integrated performance measurement of construction projects. J. Manag. Eng. **30**(6), 04014027 (2014)
35. S.S. Fanaei, S. Alkass, O. Moselhi, Performance prediction of construction projects using soft computing methods. Can. J. Civ. Eng. **46**, 609 (2019)
36. L. Knapcikova, D. Perakovic, Innovations in ICT, management, industrial and materials engineering. Wirel. Netw **28**(2), 853–854 (2022)

Index

L. Knapčíková, D. Peraković (eds.), *8th EAI International Conference on Management of Manufacturing Systems*, EAI/Springer Innovations in Communication and Computing, https://doi.org/10.1007/978-3-031-53161-3